STATES OF OBSERVANCE

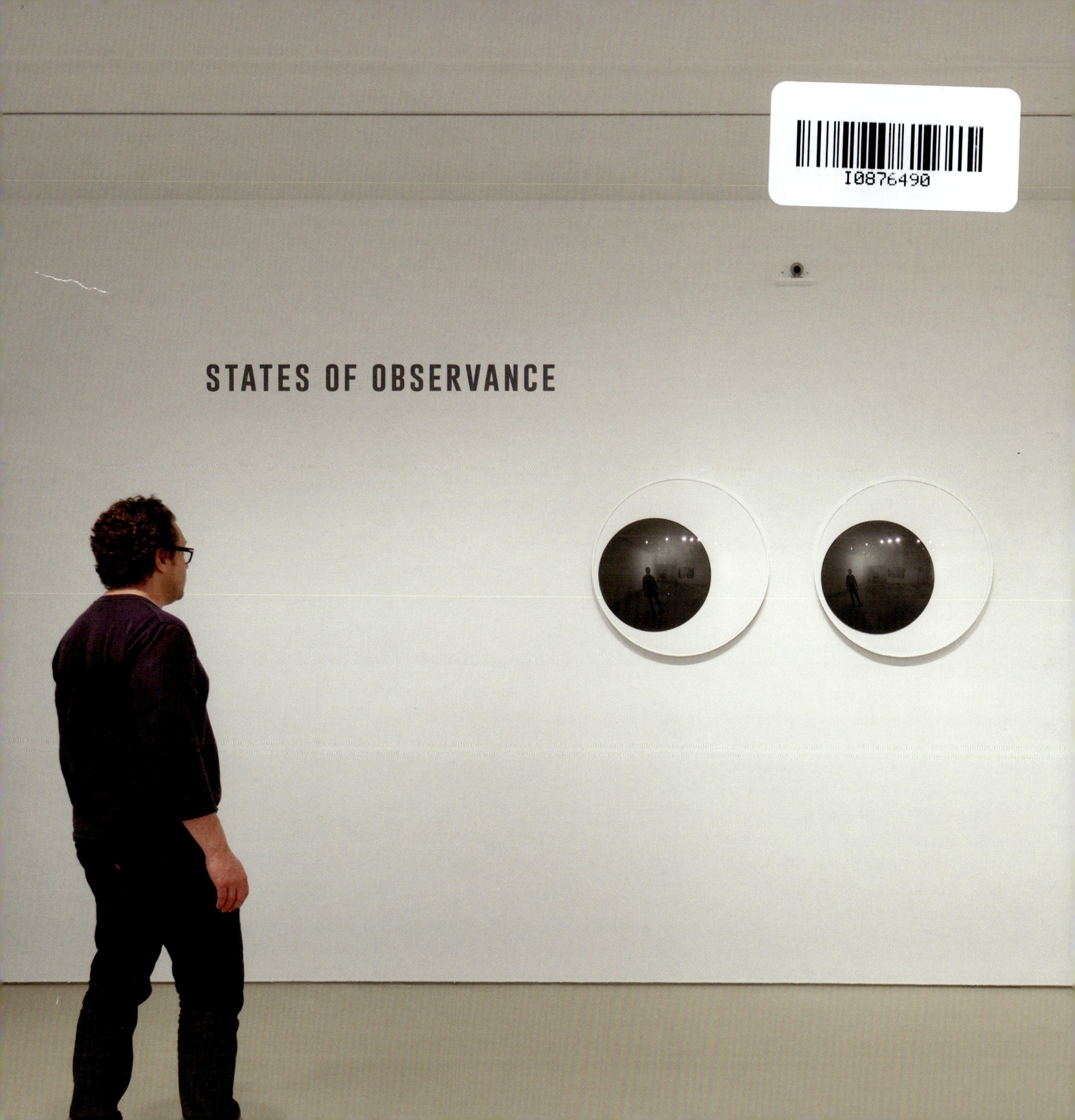

McGill-Queen's/Beaverbrook Canadian Foundation Studies in Art History

Martha Langford and Erin Silver,
series editors

Recognizing the need for a better understanding of Canada's artistic culture both at home and abroad, the Beaverbrook Canadian Foundation, through its generous support, makes possible the publication of innovative books that advance our understanding of Canadian art and Canada's visual and material culture. This series supports and stimulates such scholarship through the publication of original and rigorous peer-reviewed books that make significant contributions to the subject. We welcome submissions from Canadian and international scholars for book-length projects on historical and contemporary Canadian art and visual and material culture, including Native and Inuit art, architecture, photography, craft, design, and museum studies. Studies by Canadian scholars on non-Canadian themes will also be considered.

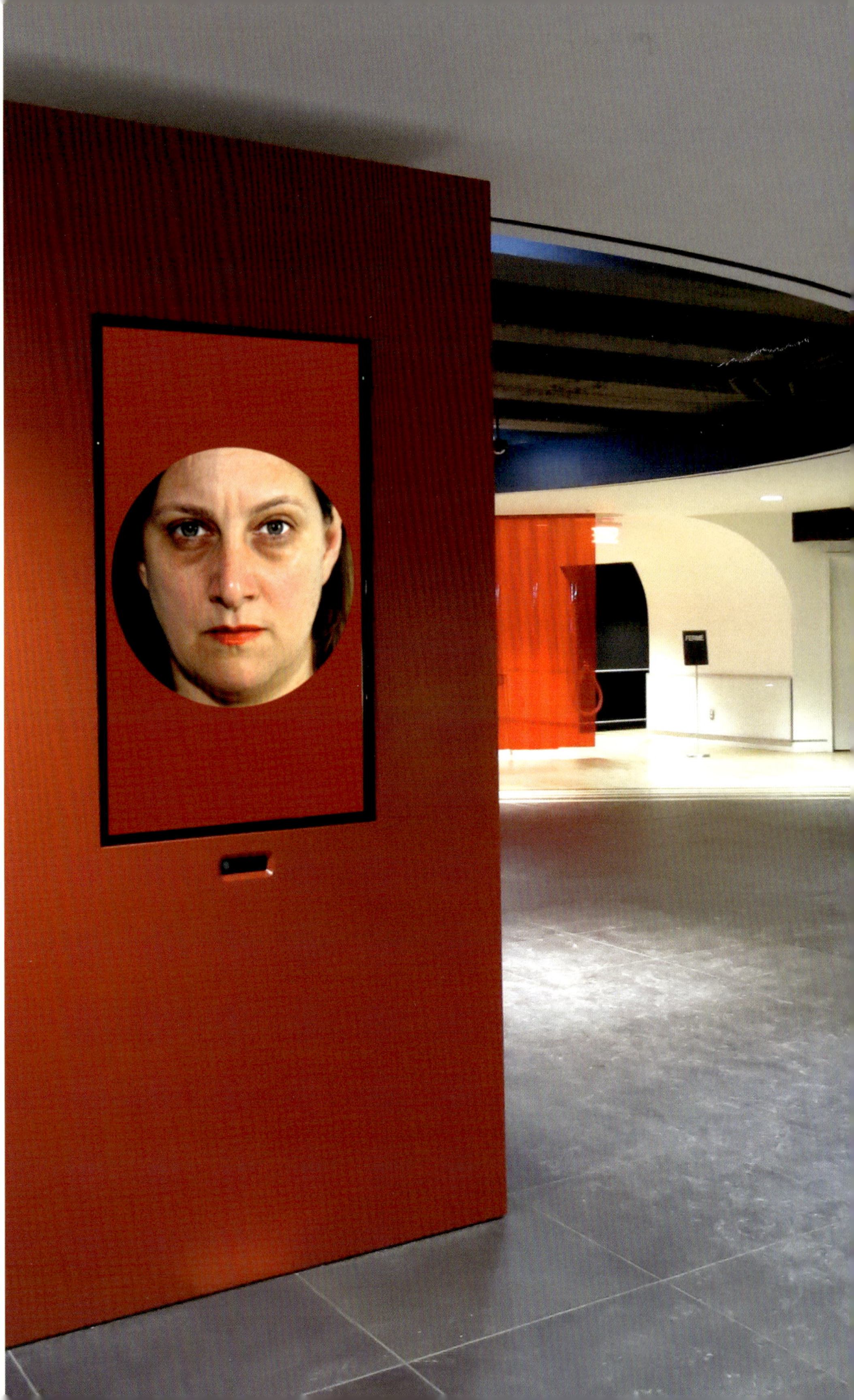

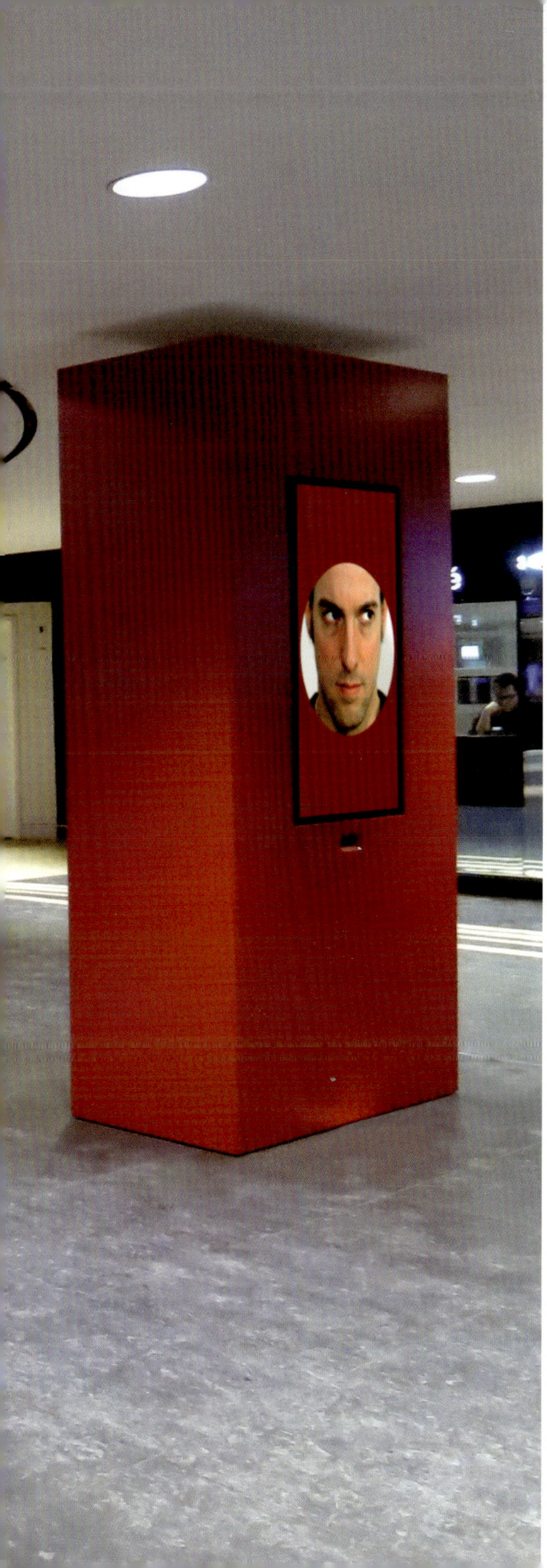

STATES OF OBSERVANCE

Contemporary Art and Surveillance Logic in Canada

SUSAN CAHILL

McGill-Queen's University Press
Montreal & Kingston

ISBN 978-0-2280-2897-0 (paper)
ISBN 978-0-2280-2980-9 (ePDF)
ISBN 978-0-2280-2927-4 (ePUB)

Legal deposit fourth quarter 2026
Bibliothèque et Archives nationales du Québec

Printed in Canada on acid-free paper.

This book has been published with the help of a grant from the Federation for the Humanities and Social Sciences, through the Scholarly Book Awards, using funds provided by the Social Sciences and Humanities Research Council of Canada.

Funded by the Government of Canada | Financé par le gouvernement du Canada | Canada

Canada Council for the Arts | Conseil des arts du Canada

We acknowledge the support of the Canada Council for the Arts.
Nous remercions le Conseil des arts du Canada de son soutien.

McGill-Queen's University Press in Montreal is on land which long served as a site of meeting and exchange amongst Indigenous Peoples, including the Haudenosaunee and Anishinabeg nations. In Kingston it is situated on the territory of the Haudenosaunee and Anishinaabek. We acknowledge and thank the diverse Indigenous Peoples whose footsteps have marked these territories on which peoples of the world now gather.

Library and Archives Canada Cataloguing in Publication

Title: States of observance: contemporary art and surveillance logic in Canada / Susan Cahill.
Names: Cahill, Susan (Lecturer in art history), author.
Series: McGill-Queen's/Beaverbrook Canadian Foundation studies in art history.
Description: Series statement: McGill-Queen's/Beaverbrook Canadian Foundation studies in art history | Includes bibliographical references and index.
Identifiers: Canadiana (print) 2026015377X | Canadiana (ebook) 2026015380X | ISBN 9780228028970 (paper) | ISBN 9780228029809 (ePDF) | ISBN 9780228029274 (ePUB)
Subjects: LCSH: Art, Canadian—21st century. | LCSH: Surveillance in art. | LCSH: Settler colonialism—Canada.
Classification: LCC N6545.6.C34 2026 | DDC 704.9/493234482—dc23

This book was designed and typeset by Lara Minja in Chapparal Pro 10.5/15pt.
Copyediting by Barbara Tessman.

McGill-Queen's University Press
Suite 1720, 1010 Sherbrooke St West, Montreal, QC, H3A 2R7

Authorized safety representative in the EU: Mare Nostrum Group BV, Doelen 72, 4831 GR Breda, the Netherlands, gpsr@mare-nostrum.co.uk

For Mary Angela and Dorothy James

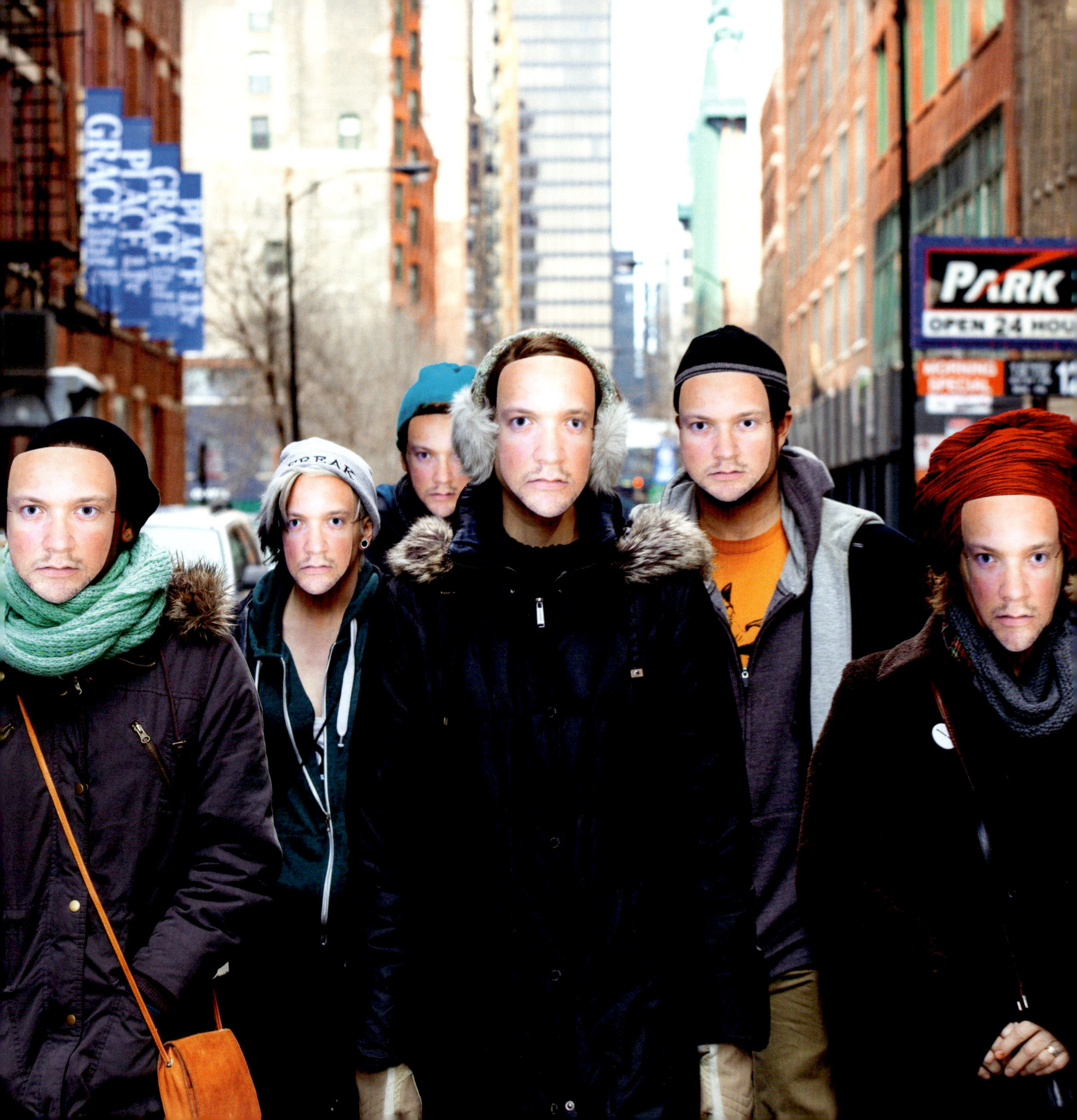
GRACE
PLACE
PARK
OPEN 24 HOU
MORNING SPECIAL

CONTENTS

FIGURES

ACKNOWLEDGMENTS

This book would not have been possible without the engaging and thought-provoking artworks that formed its foundation. These works opened up vital ways for me to explore broader cultural, political, and theoretical contexts. I'm deeply grateful to the artists for their creativity, their powerful contributions, and their generosity in allowing their work to be reproduced here. Thank you to Hiba Ali, Pat Badani, Sharlene Bamboat, Daniel Borins, Philip Cheung, Derek Curry, Heather Dewey-Hagborg, Derek Dunlop, Aljumaine Gayle, Steve Giasson, Jennifer Gradecki, Ayesha Hameed, Antonia Hirsch, Edwin Janzen, Ursula Johnson, Dave Kemp, Thomas Kneubühler, Michael Lewis, Jennifer Marman, Ricarda McDonald, Stéfy McKnight, Kate McQuillen, Alexis Mitchell, Wanda Nanibush, Abdi Osman, Desiree Palmen (with Danniel Acosta, Laticia Baptiste, Jaden Jugovic, Carson Pechawis, Sage Markham, Dene Couture, and Raven Kahnapace-Bird), Jon Rafman, Genevieve Robertson, David Rokeby, Anita Schoepp, Leonardo Selvaggio, David Spriggs, Donna Szoke, Shaheer Tarar, Pierre Tremblay, Syrus Marcus Ware, and Jay White. The work of this book is informed and guided by the compelling and generative creative practices of these artists.

I would also like to thank the many curators and cultural practitioners whose group exhibitions offered thoughtful and inspiring models for engaging with multiple artworks across diverse themes. In particular, curators such as Jan Allen, Srimoyee Mitra, Philip Leonard Ocampo, Sophia Oppel, Sarah EK Smith, and Su-Ying Strang, among others, curated exhibitions on the theme of surveillance that meaningfully shaped my own understanding of how to navigate and interpret artworks in relation to one another and within broader contexts.

A number of people and different institutions provided supports that helped bring this project to life. Thank you to the team from McGill-Queen's University Press: Jonathan Crago, Kathleen Fraser, Lisa Aitken, Catherine Bienvenu, Paloma Friedman, Filomena Falocco, and Barbara Tessman. Thank you to the people and residencies that created some space for me to write and reflect on this work: Jane Walker at Union House Arts; James Ellis and Sean Lindsay at the Calgary Institute for Humanities (University of Calgary); Kinga Olszewska, Amelia Kiddle, and the Dean's Office in the Faculty of Arts for the Faculty of Arts Dean's Writing Retreat (University of Calgary); and Mary-Anne McTrowe and Shawn Coburn at the Gushul Studio Writer Residency Program (University of Lethbridge). Thank you to the Social Sciences and Humanities Research Council and the Centre for Research in the Fine

Arts for funding that helped me explore and workshop the research that developed into this book. Thank you to the Universities Art Association of Canada and the Surveillance Studies Network for organizing conferences and for creating networks of scholars that help nurture the fields of art history and surveillance studies. And a heartfelt thank you to Taryn Sirove, whose thoughtful insights, generous engagement, and sharp editorial eye were invaluable during the editing stages of this project. Your care and attention not only strengthened the text but also made the process a much more collaborative and rewarding experience.

There are many colleagues and students whose support and guidance have been deeply meaningful throughout the development of this research. I'm especially grateful to Philip Hiscock, who offered one of my earliest models of what it means to be a generous and thoughtful mentor. Thank you to Janice Helland, whose example of professionalism, leadership, and care continues to inspire the way I approach my own work and relationships. I've also been fortunate to work with a number of dedicated graduate and undergraduate students at both Nipissing University and the University of Calgary, whose curiosity, insight, and research support contributed in vital ways at different stages of this project. Thank you to Morgan Campbell, Johanna Fraser, Sarah Marinelli, and Katherine Roberts for your engaged contributions. I've also found a small but mighty community in surveillance studies colleagues such as Julia Chan, Stéfy McKnight, Torin Monahan, and sava saheli singh – scholars and practitioners whose conversations, collaboration, and camaraderie have helped shape and sustain my thinking around surveillance and creative practice. I'm grateful for your presence and your work.

In each of the places I've called home, I've been fortunate to find generous, thoughtful people whose professional and personal support helped build a sense of community and belonging. My heartfelt thanks go to those who, at different moments, extended kindness, solidarity, encouragement, and friendship. I'm especially grateful to Judy Anderson, Amanda Balsys, Amanda Burk, Alison Coultas, Erin Dokis, Mireille Eagan, Hilary Earl, Anuradha Gobin, Christine Grandy, Natalie Grynpas, Stephen Guy, Sara Halley, Dean Hay, Jacqueline Jenkins, Nicole Jensen, Leah Kenyon-George, Serena Kataoka, Jessalynn Keller, Heather Leier, Gillian McCann, Annie Rudd, Ashley Scarlett, Dona Schwartz, Katrina Srigley, Trevor Stark, Erin Sutherland, and Erin Talbot.

Thank you to my family, whose care and love have come in many unexpected yet deeply appreciated ways. To my mom and dad, thank you for your generous hospitality and support, and for providing a safe space with childcare during the earliest and most uncertain days of the COVID-19 pandemic. Andy, I'm grateful to you for your kindness and unconditional friendship. To Nestar, thank you for the many years of care and companionship we shared. I deeply appreciate your ongoing commitment as a supportive co-parent. To my most beloved Dorothy: you fill our home with love, joy, curiosity, and a spirit of adventure, and I'm endlessly grateful to share in your journey and be part of your life. And to Birdie: good dog.

And finally, to my dear friends: Deanne Bambrick, Jeff Barbeau, Stacey Brown, Dia Da Costa, Mél Hogan, Kristy

Holmes, Brianne Howard, Julie Kiernan, Erin Morton, and Carla Taunton. Thank you for your unwavering presence, your deep generosity, and the laughter, honesty, and care you bring into my life. Your friendships have been a constant source of strength, joy, and grounding throughout this journey. All of you have stepped forward in moments when life's obstacles felt insurmountable, and I am endlessly grateful. I am better – always – for having you in my life.

STATES OF OBSERVANCE

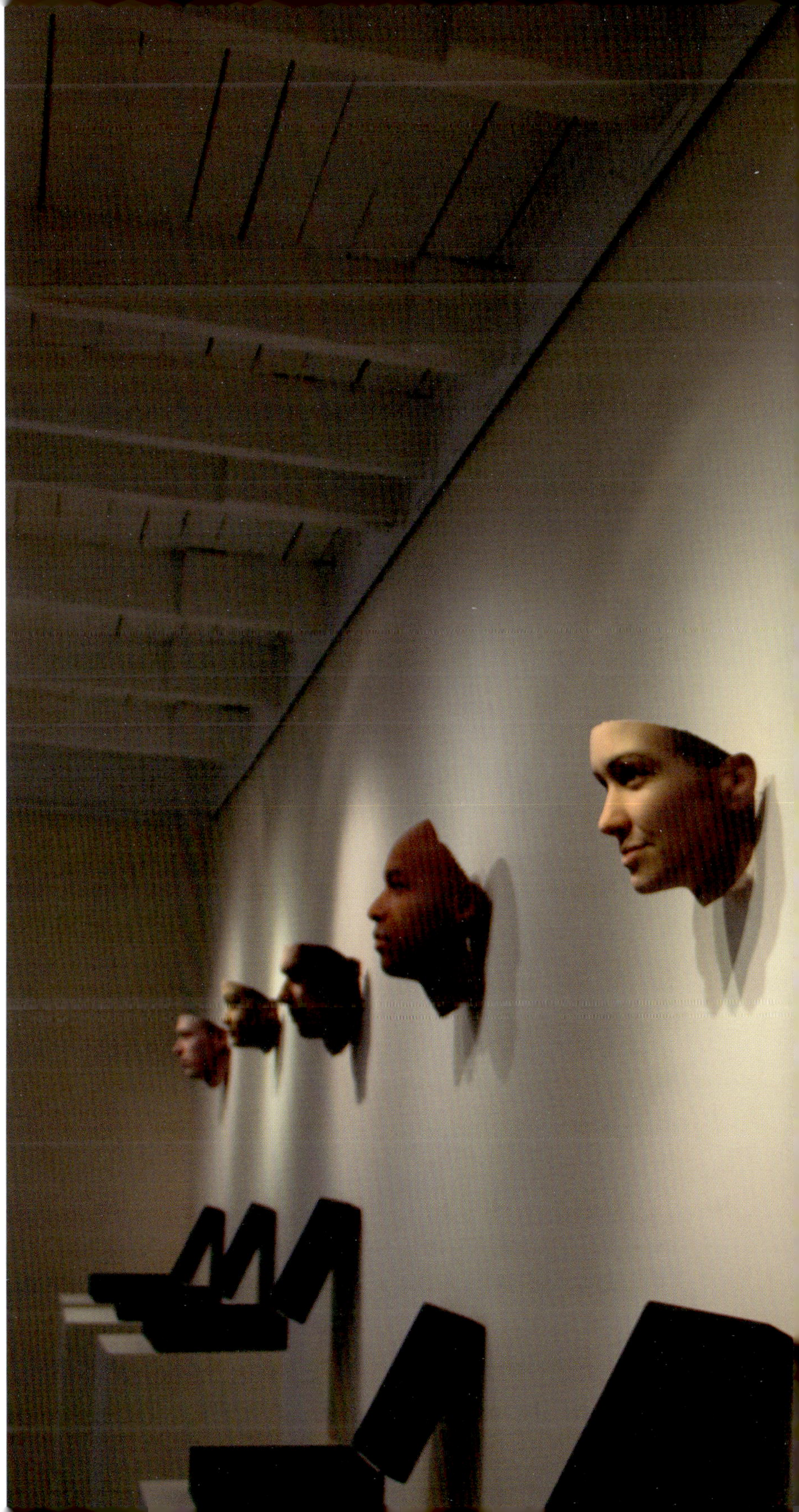

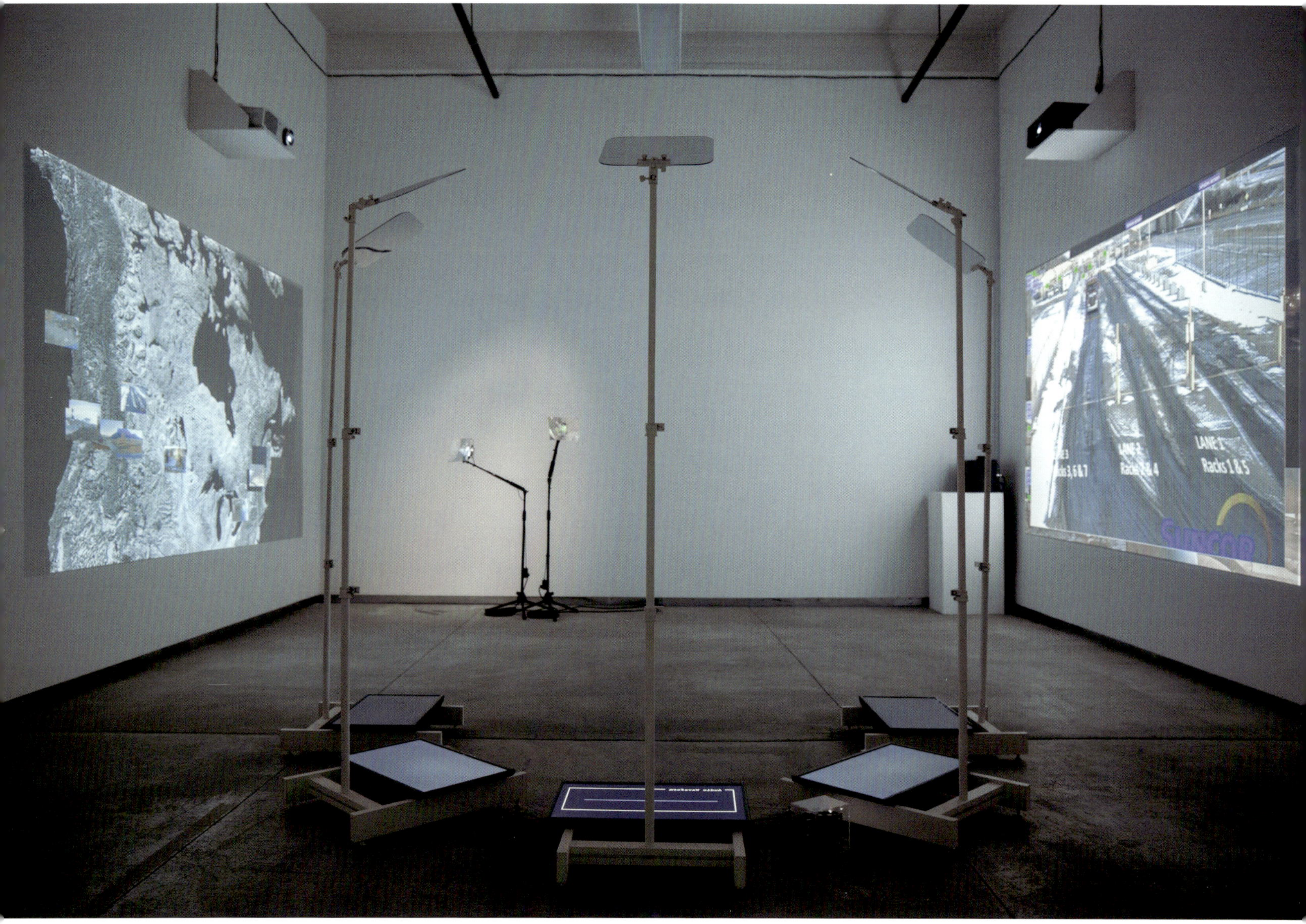

I.1 • Shaheer Tarar, *Jack Pine*, 2019.

INTRODUCTION

There is no form of surveillance that is innocent.

Lisa Nakamura, "Afterword"

Since encountering Shaheer Tarar's installation *Jack Pine* (2019, fig. I.1), I have returned to it multiple times.[1] This artwork projects two images on opposite walls: one is a live feed from unsecured surveillance cameras in different locations around Canada; the other is a satellite image of the country on which photos marking the geographical locales of each of the cameras are superimposed.[2] The surveillance live feeds that Tarar was able to connect to through open-access internet sites are processed through artificial intelligence (AI) image-recognition software, which misidentifies the objects, people, and landscapes in the images as often as it correctly discerns them. The artwork was originally included in the group show *Scaffolds I can no longer see*, which was curated by Sophia Oppel and Philip Leonard Ocampo and mounted at the non-profit art gallery and production space InterAccess in Toronto, Ontario, from November to December 2019.[3] When I first came across this work, I was taken by the title and its deliberate reference to *The Jack Pine* (1916–17), an oil painting by early twentieth-century Canadian artist Tom Thomson (fig. I.2). The Thomson work is arguably one of the best-known images in the history of Canadian art. Depicting a single pine tree in the foreground of a wilderness landscape at sunset, Thomson's painting affirms white settler imaginings of uninhabited landscapes available for national expansion and colonial projects. Considered the grandfather of modern Canadian art, Thomson became prominent through his depictions of the northern wilderness regions of Ontario and helped establish an aesthetic style that was later made internationally famous by the Group of Seven.

Although Tarar's *Jack Pine* bears very little aesthetic resemblance to its namesake, his choice to intentionally title his piece after such a prominent painting lends his own work a particular interpretative context. While different in medium and design, Tarar's artwork draws on the well-known nationalistic sentiment tied to "empty" Canadian landscapes that Thomson's painting typifies. In Thomson's painting, the landscape is rendered through an authoritative gaze that symbolizes a white settler claim to the rugged wilderness as owned, known, and tamed. Tarar, in contrast, exposes the logic underscoring such depictions, and its continuinty in relation to contemporary surveillance operations, by using geographic coordinates of surveillance cameras to plot them on a sattelite image of Canadian settler territory. Tarar himself articulated that the vision for this piece was to think about "a new Canadian landscape" that critically brings to the forefront a "specific colonial lens."[4] With this work, Tarar makes explicit connections between territories of contemporary surveillance structures and settler colonial expansion. This piece is one of an increasing number of creative projects that address the themes, methods, and technologies of surveillance structures within the Canadian context between 2001 and 2021, a period dominated by the global War on Terror. I use Tarar's artwork here as the entry point into the main goal of this book, which is to examine the ways in which art reveals the expediency of surveillance in Canada during the War on Terror to continue and extend the settler colonial project.

Since the 11 September 2001 attacks on the World Trade Center and the Pentagon in the United States, the rhetoric and policies of security and surveillance have been central to the logic of the War on Terror. The increased use of surveillance technologies such as closed circuit television (CCTV), reconnaissance satellites, and unmanned aerial vehicles (UAVs) or drones, has become part of the public and private landscape in many countries, including Canada. These forms of elevated vigilance, which impose military surveillance tactics to monitor everyday behaviours, spaces of consumption, and civilian life, have been normalized as essential to national and

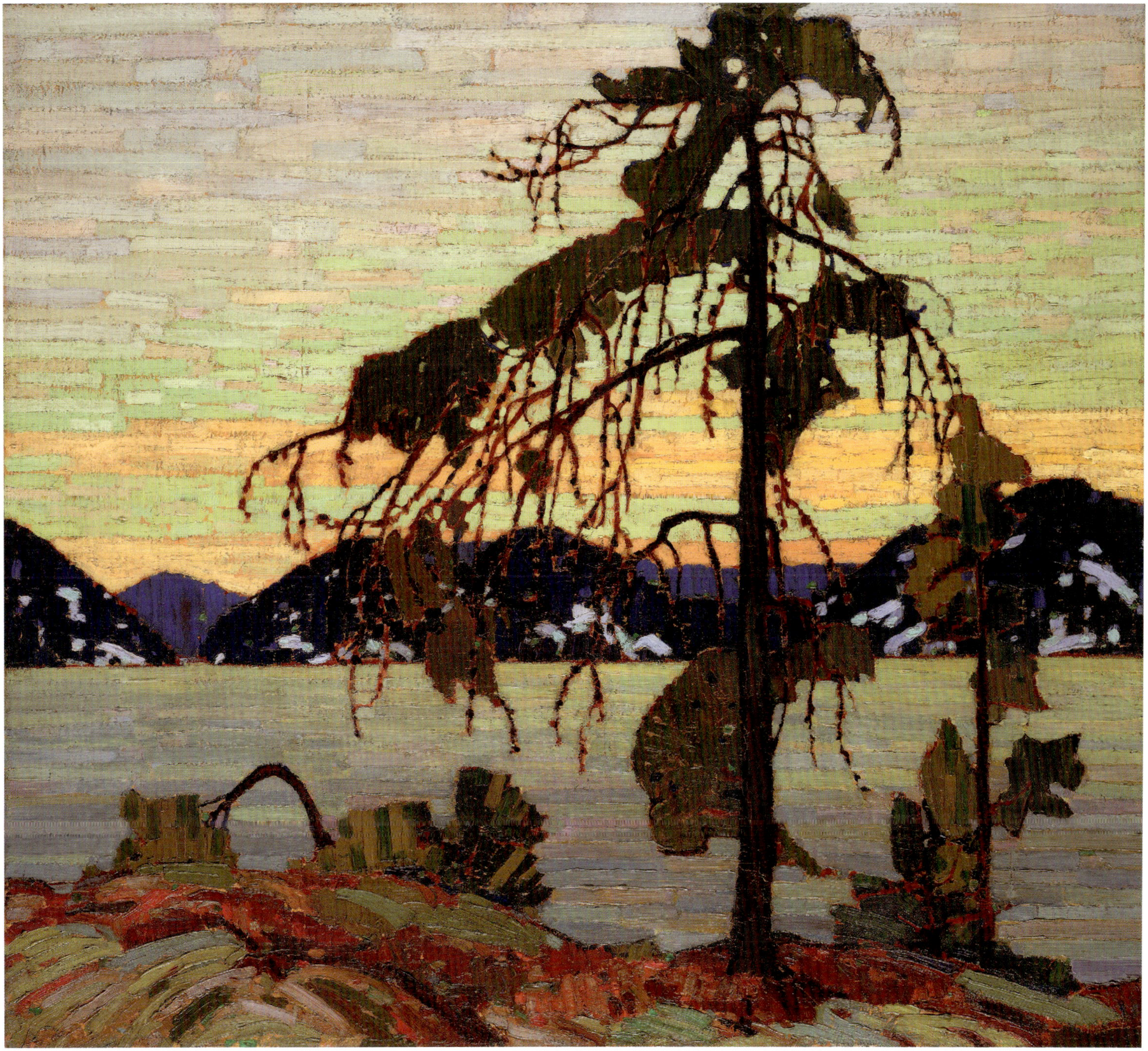

I.2 • Tom Thomson, *The Jack Pine*, 1916–17.

global security. Since the beginning of the War on Terror and the increase in use – and, significantly, wider public acknowledgment – of surveillance measures in the name of safety and security, there has been a correlative increase in the number of creative works engaging with surveillance. Artists and curators across Canada have produced a rich field of creative engagements with surveillance, and, since 2015, I have collected data on these creative projects as part of a larger research effort, the Art and Surveillance Project. The central output of the Art and Surveillance Project has been an online archive that catalogues and presents a database of creative practices and practitioners addressing surveillance systems in Canada after 2001.[5] Some of these artworks have been exhibited singularly or as part of an artist's larger body of work; others have been included in thematic group exhibitions. My goal with the Art and Surveillance Project has been to engage with the depth and breadth of creative practices dealing with contemporary surveillance in Canada. With this online database and archive, I have been interested in circulating the wide variety of works being produced, and in highlighting their significance as participants in academic and public discussions of contemporary relevance. In addition, in bringing together these varied examples, which are otherwise disconnected from one another, I have sought to begin the process of thinking through what types of meanings may be produced in their juxtaposition.

States of Observance is a critical extension of this database. This book brings the online collection to life: it offers critical and creative analysis by placing different artworks together and in relation to an interdisciplinary body of scholarship and research relevant to surveillance. The title of the book, *States of Observance*, refers to the different dimensions and contexts that enable and encourage certain forms of seeing, revealing, and knowing. The term "states" refers to a form or condition, but it also refers to governments or government-sanctioned agencies. The term "observance" can mean keeping or celebrating a ritual. It can also refer to both passive and active actions: observance as conforming or obeying, or observance as watching, noting, or perceiving. The title intends to play with expectations about who has the power to watch, collect, monitor, and police through surveillance systems, and to point to ways in which creative works can offer new ideas – and new forms of resistance and power – related to rethinking and re-evaluating these expectations.

This book focuses on creative projects produced or exhibited in Canada during the period between 2001 and 2021. I chose these years because they are defined by the rhetoric and practices associated with the War on Terror, which affirmed and intensified surveillance structures as part of national and global security measures. The year 2001 marks the plane attacks on the World Trade Center in New York and the Pentagon in Washington, events that became the watershed for an international conflict against "terrorism." The year 2021 is officially accepted as the end of the War on Terror, as the US-led mission withdrew all troops and operatives from Afghanistan, and the Taliban regained control of that country. Certainly, the nature of the rhetoric around and applications of surveillance during these twenty years was not homogeneous; what I am interested in here, however, are the commonalities and recurring trends during this time. These years encapsulate a period when surveillance systems expanded as a national instrument for guarding territorial borders, and this book examines how the War on Terror served as justification for imperial actions.

Although the 2001–21 period marks a discernible configuration in the way surveillance measures were discussed and applied, I assert throughout this book that the post-2001 period represents a continuation of a longer history of settler colonialism in Canada. I see settler colonialism, as the foundation of the Canadian nation-state, working in tandem with other systems of oppression and inequity, including capitalism, white supremacy, racism, cisheteropatriarchy, and ableism.[6] Certainly these structures operate unevenly and specifically within the broader goals of settler colonialism, yet I bring them together in this way to highlight how they are utilized to uphold, bolster, and further a particular vision of Canada. These systems are the scaffolding that uphold and reproduce the settler colonial project. So, while the central context of analysis of this book is settler colonialism, I also examine the ways in which this context is co-constituted and reinforced by these other systems. And while the artworks throughout this book do not necessarily address each form of oppression directly, I argue that they implicate mutually reinforcing, exclusionary systems of power that are interlocking and fundamentally inseparable. In so doing, I argue that these creative projects make a contribution to critical thinking about surveillance and its associated systems of power. A key aim of this book is to show how art produced during this time participates in shaping and reshaping our understanding of this historical moment within the Canadian context.

What We Are Talking About When We Talk About Surveillance Art

Artworks and exhibitions provide a unique lens through which to view the state of contemporary surveillance because they can render the familiar forms of surveillance structures unfamiliar, and thus provoke renewed understandings of them. Creative practice enables a mode of critique that is unavailable in other forms. "As a form of critical reflection, analysis, and description," artist and new media scholar Maciej Ożóg articulates, "art surveils the surveillance society and indicates its essential features."[7] Creative practices surveil the agents and systems of surveillance and present them to audiences in ways that denormalize these structures to reveal the often invisible and unquestioned logic that governs them. As communications and surveillance studies scholar Torin Monahan highlights, artworks deploy "tactics of defamiliarisation to draw critical attention to everyday surveillance that has become mundane."[8] They bring into clear view the visible and invisible surveillance mechanisms that structure and shape contemporary life. Artworks are capable of illuminating the contours and contestations of this historical moment and of contributing to its social, cultural, and political configuration. Creative practices re-present surveillance structures to reimagine and destabilize the processes, technologies, and agents that have contributed to normalizing surveillance and surveillance viewing.

There is no single accepted label or definition used to identify creative works that can be characterized as "surveillance art." Sociologist Andrea Mubi Brighenti introduces the term "artveillance" to identify "the domain of the reciprocal influences and exchanges between art and surveillance."[9] He also deploys the term "surveillance art," which he defines as "contemporary artwork that in some way hints to or deals with topics, concerns and procedures that fall within the interest of surveillance studies."[10] Theatre studies scholar Elise Morrison adopts "surveillance art" as a more pointed category, defined as "performances or installations in which surveillance technologies are central to their production, design, content, aesthetics, and/or reception."[11] Monahan introduces the concept of "critical surveillance art," which extends from "art or performance projects that either focus on surveillance problematics or incorporate surveillance technologies into their aesthetic production," to also include an intent toward resistance and criticality, as artworks "tackle issues of domination, oppression, and inequality."[12]

A variety of art, cultural, and scholarly contexts have brought together artworks under the category of surveillance art, including large-scale international exhibitions such as *CTRL [SPACE]: Rhetorics of Surveillance from Bentham to Big Brother* (2002; curated by Thomas Y. Levin), *Exposed: Voyeurism, Surveillance, and the Camera since 1870* (2010; curated by Sandra S. Phillips, Simon Baker, and Ann Coxon), and *Watched! Surveillance, Art, and Photography* (2016; curated by Louise Wolthers, Dragana Vujanovic, Niclas Östlind, and Ann-Christin Bertrand). By categorizing different creative projects in this way, these exhibitions (among others) demonstrate the way in which knowledge is produced not just in the single artwork but also through how different artworks engage in conversation with each other. The category of surveillance art becomes a pathway or shortcut to identify a point of address, topic, and theme within creative projects, and unites them with a shared point of analysis to generate larger questions and conversations around the topic of surveillance.

Yet, the category of surveillance art needs to be applied mindfully. It cannot be presumed to be self-explanatory or to summarize the motivations of a creative project. Certainly, there are characteristics and tropes that align a work with surveillance. Artists and researchers Fernanda Bruno, Paola Barreto, and Milena Szafir engage with the category of surveillance art in Latin America by thinking through the way in which artworks deploy a certain form of aesthetics, which they identify through "the appropriation of dispositifs such as closed circuit video, webcams, satellite images, algorithms and computer vision, among others, placing them within new visibility, attention and experience regimes." At the same time, while deploying surveillance art as a category, they also highlight the need to recognize this term "more as a vector of research than the determination of a field."[13] To engage with surveillance art as a critical and creative category is to analyze the way in which creative practices marry aesthetic or representative function with broader social, political, and historical functions.

What I am asserting here is that surveillance art is not simply an aesthetic category; it is a thematic mode of creative engagement. The category of surveillance art is useful in that it is a shorthand to identify a shared critical context, but we should be cautious about the extent to which we allow it to do the heavy lifting of explaining what an artwork does. Such categorizations run the risk of becoming conclusive or circumscribed by situating artworks as decorative or illustrative or representing what is already known. For media studies scholar Gary Kafer, artists engaging with surveillance contexts do not simply deploy surveillance technologies or processes but, rather, "aim to mobilize a critical discourse on the rise and proliferation of surveillance in contemporary society by engaging with the politics and power relations configured by surveillance systems."[14] When we use a category like surveillance art, it needs to be a stepping stone to generating a conversation about contexts of surveillance by thinking through creative practices as critical contributions in and of themselves that produce and generate ways of being and understanding the world.

Artworks produce meanings both as individual works that engage with topics of surveillance and in relation to each other as part of larger conversations about surveillance. Gathered together in different arrangements, art can enable audiences to see larger stories in new ways. Yet this "seeing" is not tied purely to the visual, although this is a common way to describe the contribution of creative projects. Art that engages with surveillance, in particular, is in need of being conceptualized as beyond a form of seeing. Surveillance in its relationship to seeing is apparent not only in terms of visibility and recognizable visual forms but as a matter of visuality itself.[15] Capturing how social conditions are constituted through visual discourse, the concept of "visuality" refers to the ways in which different systems become visually knowable. As visual culture theorist Nicholas Mirzoeff writes, "visuality is very much to do with picturing and nothing to do with vision, if by vision we understand how an individual person registers visual sensory impressions."[16] Visuality works in the construction of surveillance systems and their logic via animating social conditions through their representation. The idea of seeing here reaches beyond a simple matter of visual understanding and presentation. It is seeing in the sense of knowing: I see – I understand – I have access to this knowledge.

Although discussions around the visuality of surveillance often take the visual aesthetics of surveillance as their entry point, arguably because the visual form of surveillance is the most immediately recognizable (think top-down camera angles, drone imagery, CCTV footage, and so on), the visuality of surveillance is never purely visual. The type of seeing associated with surveillance angles of vision does not simply replicate the recognizable aesthetic of popular surveillance references; rather, surveillance visuality accounts for, as visual and media culture scholar W.J.T. Mitchell writes, how "the visual construction of the social field" of surveillance constitutes and challenges that field, and can function in embodied and material ways.[17] As film studies scholar Rémi Lauvin asserts, "the visibility at stake, here, is not purely optical."[18] Rather, the seeing of surveillance works in partnership with the other senses as part of the process of generating access, knowledge, control, and understanding.

What I argue here is that creative practices contribute to and can challenge this field of surveillance visualities and thus the way social conditions are known. They both reproduce and constitute ways of thinking about systems of surveillance and their role in organizing contemporary social and political meanings. Just as media culture scholar Nicole Fleetwood explains with respect to representations of Blackness, "cultural practices and products reveal information about how we come to know what we know."[19] Artworks that deploy, critique, play with, and adapt modes of surveillance seeing do not simply replicate established aesthetics and forms of surveillance that exist generally; they also constitute new ways of contemplating surveillance and its forms of information gathering. The proliferation of surveillance as a recognizable form of looking and knowing, as visual communications scholar Jonathan Finn describes, "has become a key feature of contemporary visual culture and of life more generally."[20]

Artworks generate an experience that is distinct from the material realities of surveillance structures but is discursively and politically crucial in the ways they reveal the beliefs, expectations, perceptions, and rationale of why, when, and for whom surveillance operates. Certainly, the notion of art as experiential is not new. For example, in the early twentieth century, the American philosopher and psychologist John Dewey asserted the need, for those undertaking "to write upon the philosophy of the fine arts," to ensure "continuity between the refined and intensified forms of experience that are works of art and everyday events, doings, and sufferings that are universally recognized to constitute experience."[21] What we are talking about when we talk about "surveillance art," then, is less than a category defined by the visual re-presentation of surveillance-related forms, and more a type of creative experience that forces an encounter between art project and audience to encourage a form of engagement and critical awareness of surveillance contexts. The point of talking about art in terms of experience, rather than as a static artistic category, is to ensure the intent and analysis continually move toward cultural critique, rather than simply generate a list of what characteristics of surveillance the artwork integrates. This approach places artworks in the context of visuality, insofar as the creative practices reveal the construction of surveillance structures and, often, bring into view the specific reasonings and logics that govern them.

Seen in this way, art becomes a form of political engagement, a mode of understanding, addressing, critiquing, and reframing the socio-historical context of surveillance that can reimagine the political ground on which particular ideas – such as those tied to nationhood, belonging, citizenship, uncertainty, terror, security, and privacy – are based. Art is a manner of doing politics. In other words, art does not simply reflect its contexts but, rather, actively generates meanings and understandings of the world. Creative works are conversant in ongoing discussions with which their content and subject matter engage, as they bring forward their own ideas as well as develop meanings in dialogue with one another.

What I Am Talking About When I Talk About Surveillance Art

All of the artworks included in this volume come together in their creative and critical address of surveillance contexts. But I also applied specific criteria to guide the selection of particular creative works within the context of this project. These criteria include geographic and temporal elements: specifically, the art projects must have been circulated in Canada between 2001 and 2021. While some of these works were originally produced or exhibited outside Canada, their eventual inclusion in shows and public presentations within the Canadian cultural space contributes to national conversations and perceptions of surveillance. The engagement between these works and local audiences functions as part of a national conversation about surveillance and the War on Terror, while contributing to broader cultural understandings of these contexts. In other words, the Canadian context of their exhibition drives the interpretation of these works, and the works themselves participate in shaping the Canadian context. Their inclusion here is intended to help us think through the types of cultural conversations that are happening in relation to surveillance and the nation-state, but it also acknowledges that the boundaries of the nation are not static: they work in reciprocity with international discussions.

My selection criteria also extend to particular methods and approaches to the topic: the creative works must engage aspects related to the broad applications of surveillance structures; use the technologies of surveillance within the piece; implement modes of viewing and being that replicate surveillance structures; or illuminate how information is shared or monitored, and to what ends. To be sure, in the research and development of this book, these criteria were moving targets. It is difficult to establish hard and fast rules to define and classify all contemporary creative practices addressing the topic of surveillance, but I offer these criteria as a start. Works have been and continue to be produced that push the parameters I outline here, a reflection of the evolving and somewhat indefinable creativity of cultural practitioners as they address different ways of being in the world in the context of contemporary surveillance in Canada.

Another key element I require of the artworks in *States of Observance* is that they are clearly situated in relation to surveillance practices: I have been careful not to grab and draw works into the context of art and surveillance in Canada if that discussion has not already existed around the work. So the artworks included here have been previously contextualized within topics of contemporary surveillance, either by the artists themselves or through their inclusion in thematic group exhibitions or in published interviews or art reviews. I am cognizant of the issues with excavating the meaning of creative practices solely for my own purposes; that is, I have tried to avoid including an artwork simply because I felt I could make it work with the analyses I wanted to put forward. In this context, I have taken particular care with respect to creative works produced by artists with positionalities and from communities outside my own. As a white settler, cisgender, non-disabled woman, I am aware of perspectives, experiences, and circumstances of being in the world that I can and cannot speak to. I am also aware of the historical and ongoing tendency of academia – and the histories of whiteness and extraction within it – to pilfer knowledges, histories, and ideas from Black, Indigenous, racialized diasporic, 2SLGBTQIA+, and disabled people, scholars, and communities, and to repurpose them for its own gain. Social work practitioners and scholars Andrea Murray-Lichtman and Siham Elkassem identify the tendency of white academics to enact an academic voyeurism of the white gaze, a way of engaging in scholarship that is tethered to "imperialist White supremacist patriarchal capitalism." Such voyeurism extracts and interprets the work and lives of racialized folks to these scholars' own ends and then "perpetuates the (in)visibility of Black, Indigenous, and People of Colour (BIPOC) scholars and communities."[22]

Certainly, I am aware of the historical and ongoing dominance of the gaze of academia, and I am sensitive to how I bring analyses to the artworks throughout this book. The work of this book is to put artworks engaging with surveillance in conversations with scholarship from the field of surveillance studies and critical readings of Canada as a settler colonial project. While the artworks themselves may not address contexts of settler colonialism directly, I take as motivation their address of surveillance systems as a way to understand surveillance as a logic that directly participates in shaping the settler colonial project. As much as possible in my discussions of creative projects, I use the artists' own words from interviews, articles, or exhibition notes to contextualize their work. Throughout my analyses and discussions of the many different artworks in this book, my intention is not to speak for the artists, but rather to speak to the works and knowledges they bring forward in relation to the other works of art and themes I have identified. This awareness of trying to speak *with* but not *for* is critical, given that the topic at hand is surveillance – an issue long relevant to people and communities inequitably targeted based on race, gender, class, sexuality, religion, and ability. After the start of the War on Terror, surveillance contexts became newly visible to people like me who, because of race and class privilege in Canada, have often been ignorant of the role of surveillance in settler colonialism. After all, it is often one's own experiences of surveillance systems that dictate the extent to which one can recognize and name them. I aim to bring sensitivity to my approach in both selecting the works and apprehending how surveillance as a topic is addressed within them.

In addition to addressing different aspects of Canadian surveillance between 2001 and 2021, all of the pieces discussed here function as fine art. While a variety of movies, documentaries, journalism, and photography attend to the contemporary context of surveillance inside and outside Canada – and there is certainly much work to be done in this broader field of visual culture – I invite in as case studies only those works that were produced with the intention that they be considered creative fine art practices. "Fine art," for my purposes, refers to works that have been produced by those who self-identify as artists or who claim a certain artistic intent with their projects, as well as works that are circulated in the particular context of galleries, museums, artist-run centres, and public art sites. In this book, the concept of "creativity" itself functions as a mode of address and an approach to making that is central to the outcome and interpretation of a piece. Certainly, this focus does not dictate a work's form or medium, nor does it narrow its approach to traditional modes of artistic practice; indeed, the works included in the following chapters illuminate the widely varied approaches that characterize contemporary art practice. My specific interest in fine art gestures to the ways in which art produces unique sites of inquiry and opens up possibilities of the imagination, offering new and innovative perspectives to ongoing conversations.

Surveillance and Its Logic

Many scholarly definitions conceptualize surveillance, particularly in the field of surveillance studies. Media studies scholar Mark Andrejevic distills these definitions down to the main idea of surveillance as "the coupling of information collection and use with power."[23] Surveillance studies scholar David Lyon has an oft-cited definition of surveillance as "any collection and processing of personal data, whether identifiable or not, for the purposes of influencing or managing those whose data has been garnered."[24] Sociologists Kevin Haggerty and Richard Ericson summarize surveillance as the "collection and analysis of information about populations in order to govern their activities."[25] The scholars involved in The New Transparency Project define surveillance as "any systematic focus on personal information in order to influence, manage, entitle, or control those persons whose information is collected."[26] In his recent book on race and the cultural production of surveillance, Monahan defines surveillance as "a principal mode of governance deployed by institutions and individuals, through both technical and nontechnical means, to assert control over domains and the bodies within them."[27]

Broadly, then, surveillance comprises a combination of data gathering and control. Surveillance is attached to a wide number of monitoring activities that intend to make visible, govern, regulate, and police individuals and populations in order to create and maintain national, corporate, and institutional goals. Specifically for the purposes of *States of Observance*, I emphasize a conception of surveillance as the will and capacity to monitor and dictate those who are free to move, settle, and live, and those who are not.[28] This description certainly works in conversation with the definitions presented above, as well as others found throughout the field of surveillance studies and beyond, yet it focuses on the specificities of surveillance as a belief system particular to the Canadian context. In particular, it emphasizes the role that the broader surveillance apparatus plays in producing visibilities in order to maintain a perceived public good, which, in Canada, is characterized by a colonial imaginary tied to the control and mastery of the land and territory.

Surveillance's relationship to colonial settlement within Canada is a key point of analysis that is necessary to understanding surveillance systems and their logic as part of the national project. Legal historian Neil Nunn describes settler colonialism as "a territorial project that is centred on the accumulation and control of dispossessed land."[29] Surveillance's role in governing movement, settlement, and life is central to its understanding, deployment, and applications in Canada. In this context, surveillance in its relationship to the national history of the Canadian state is inextricably linked with longer histories of and attachments to settler colonialism. In this conceptualization, I draw on historian Patrick Wolfe's formulation of settler colonial invasion and occupation as a structure rather than an event.[30] Surveillance in Canada is part of this structure. Surveillance is not a singular thing or inert tool; rather, it encompasses the logic or beliefs that underlie its ongoing applications. Surveillance, then, is not just about what, how, and where systems of monitoring and data gathering operate; importantly, it represents the underlying belief system – the logics – that normalize and justify these systems in the first place.

In effect, surveillance can be considered a shorthand for the "power asymmetries, by relations of domination and extraction, that are often obscured by the technical apparatuses being used," as Monahan articulates.[31] In the Canadian context, these "power asymmetries" take their form as settler colonialism, which is bolstered by and anchored in the belief systems of capitalism, imperialism, cisheteropatriarchy, racism, and white supremacy. Surveillance operations relate to the specific conditions of historical and ongoing settler colonialism and its interconnections with other systems of oppression and inequities. These structures underlie and justify surveillance as a logic, which is inextricably linked to, as Indigenous studies scholar Glen Coulthard writes, "the constellation of power relations that sustain colonial patterns of behavior, structures, and relationships."[32] Surveillance facilitates the power asymmetries of oppressive systems through normalizing modes of looking, control, and authority that characterize what philosopher Gilles Deleuze conceptualizes as "societies of control."[33] Surveillance viewing represents a particular kind of access, a form of wielding and materializing the systems of power and control that produce, structure, and maintain Canada as a settler colonial state. Essentially, the premise that underlies and connects my analysis here is that *surveillance viewing is settler colonial viewing.*

In this premise, I follow the work of critical Muslim studies scholar Anouar Majid, who traces the birth of surveillance as we generally understand it today to major world events of 1492: the Spanish conquest of Granada, which was the last Muslim stronghold in Europe; the Spanish Inquisition and the expulsion of the Jewish people from Spain; and Christopher Columbus's expedition to the Americas and the subsequent violent takeover of these lands by colonial powers.[34] Each of these events is connected not only temporally but also ideologically, in that they are characterized by claims to power anchored in policing, surveilling, and containing certain people and populations. They are also connected by an underlying allegiance to cultural and racial "purity" and the racist belief that white Europeans were an inevitably dominant force. For Majid, then, surveillance as a practice developed in conjunction with colonial imaginaries and violence, whereby the policing, containment, removal, and subjugation of Indigenous and racialized people enabled territorial occupation and excavation of the land by white Europeans. That is to say, surveillance is intrinsic to the colonial process and histories of white supremacy. Thus, surveillance is not just the technologies and policies but also, and importantly, a logic – an underlying and pervasive set of beliefs that justify and rationalize its own structures.

Applications of surveillance in this sense are a materialization of "surveillance logic." Such a conceptualization helps illustrate that while the forms and applications and even cultural understandings of surveillance may shift over time, the foundational principles of surveillance systems – of looking, uncovering, monitoring, policing, and controlling, throughout historical periods and geopolitical locales – continue to inform ongoing belief systems. I highlight this history to emphasize that, although the use of surveillance systems extended and became public in new ways after 2001, the logic of surveillance is not new to the post-9/11 period. Critical race and gender studies scholars Shaista Patel, Ghaida Moussa, and Nishant Upadhyay have rightly argued that 9/11 and its aftermath are not exceptional: they represent the legacy and continuation of the historical violence of the settler colonial nation-state.[35] In other words, contemporary forms of surveillance and containment of bodies and populations in the War on Terror function within a longer genealogy of classifying and confining humanity, and of enabling extraction under colonialism and imperialism. The surveillance logic particular to the Canadian context is inextricably linked to imperial expansions and settler colonialism. This means that, while the logic of surveillance took specific forms with material realities and consequences attributable to the War on Terror, surveillance as a vehicle of colonial imaginaries and power anchored in policing, managing, and containing certain people and populations extends far into the national past.

Canada and Its War on Terror

The War on Terror relies on and takes up surveillance logic, channelling and deploying it in new and expanded ways that, according to the scholars working in The New Transparency Project, are related to a variety of factors, including "technology, law, politics, economy, culture, and our own perceptions and practices."[36] In this way, surveillance systems since 2001 have extended pre-existing modes of control, oppression, and coloniality, offering new forms of colonialisms that activate the logic and rhetoric of terror and security but deny the uneven and unequal ways in which these forms are applied and operate. Contemporary justifications for such systems rely on national narratives of the logics of surveillance and marginalization in the name of public safety and the colonial project that justified historical actions, such as the Indigenous reserve systems, which was first implemented in 1876, the Chinese Immigration Act (1885), and the internment of Japanese Canadians during the Second World War. These exclusionary actions taken by the state targeted different communities in different ways, but all relied on the belief systems of settler colonialism, which elevates and affirms the right of white settlers to life on the land above that of all others, and utilizes surveillance structures to maintain this order. The War on Terror is another historical period during which the settler colonial project in Canada utilized surveillance as a vehicle in overt ways, demanding public participation and compliance for securing public safety and national security.

Certainly, these new forms of colonialism also operate outside Canadian borders as part of the US-led international alliance of states that came together to fight against al Qaeda, the group that claimed responsibility for the plane attacks of 9/11, and to invade Afghanistan, which was targeted because its Taliban government was accused of harbouring those responsible. As a settler nation and as part of the British Commonwealth, Canada has marketed a very particular narrative in relation to its national conflicts and military history. Although Canada played a role in the war in Afghanistan, it avoided association with the initial US-led and UK-supported invasion "Operation Enduring Freedom," by entering later into the conflict and by promoting its influence as a supposedly peacekeeping, global middle power. Despite this peacekeeping narrative, which positions Canadian actions as disconnected from the aggressive military response of other nations, Canada remains a member of the North Atlantic Treaty Organization (NATO) and acted in Afghanistan as part of the International Security Assistance Force (ISAF), an extraterritorial policing unit. While Canada officially removed its military operations from Afghanistan in 2014, footage from the Kabul airport in August 2021 of the disastrous removal of NATO allies after the Taliban takeover of the Afghanistan government included, as international relations scholar Taylor Robertson McDonald highlights, images of many "Canadian citizens, permanent residents, military personnel, and interpreters among the thousands on the ground."[37]

Internal to the Canadian nation-state, the federal government enacted a number of laws and policies that effectively expanded the surveillance network in the name of security and the War on Terror. The most well-known – Bill C-51, The Anti-terrorism Act, 2015, which modified legislation passed in 2001 – provoked public discussions about surveillance, privacy, and civil liberties.[38] Extending the monitoring powers of the police and the Canadian Security Intelligence Service (CSIS), the bill, its critics argued, was too vague and could lead to dangerous and unlawful measures.[39] To be sure, Bill C-51 is only one of a series of bills passed between 2001 and 2021 that affected how information is gathered and shared in Canada. Other new or amended legislation that covers surveillance access includes Bill C-36, The Anti-terrorism Act, 2001 (October 2001); Bill C-7, The Public Safety Act (March 2002); Bill S-4, Digital Privacy Act (June 2014); Bill C-44, Protecting Canada from Terrorists Act (April 2015); and Bill C-59, An Act Respecting National Security Matters, 2017 (June 2019). Despite public protestations about the level of power such bills grant government officials and agencies – including a statement from the International Civil Liberties Monitoring Group criticizing the level to which Bill C-59 grants power that amounts to "military actions" with limited oversight and "no guarantee that there will be public reporting on these activities" – these bills were enacted.[40] They extend the reach of the Canadian surveillance state and work in tandem with Canadian participation in surveillance trends internationally.

In the international context, the history of Canada's role in broader intelligence-gathering activities, such as through its participation in the Five Eyes network, has been revealed through the 2013 circulation of previously unseen documents included in Edward Snowden's National Security Agency (NSA) disclosures. The Five Eyes is a global alliance founded in 1946 between "five anglophone countries and their security agencies: the US (NSA), the UK (GCHQ), Australia (ASD), Canada (CSEC) and New Zealand (GCSB)."[41] Perhaps unsurprisingly, this network continued and increased its capacities in the post-2001 global environment. As well, Canada established new relationships to surveillance operations and intergovernmental organizations (such as NATO) after the election of Prime Minister Justin Trudeau in October 2015. The previous federal government had withdrawn from NATO-led surveillance programs in 2012 and 2013, preferring private agreements on satellite surveillance and communications with the United States and Australia.[42] In contrast, at a NATO leaders summit in July 2016, Trudeau asserted his government's commitment to participate in and even lead information-gathering programs.[43] Key here is the Canadian state's continued commitment to and involvement in international programs of surveillance in the name of protecting national and global security during the War on Terror.

While the politics and policies surrounding Canadian domestic and international security activities provide the larger framework within which I situate my critical analysis of creative practices, I do not focus on the specifics of Canada's on-the-ground participation in the global War on Terror, or how government officials, politicians, and military operatives presented this participation as part of shaping Canadian national identity for an international audience. Rather, I am interested in how the rhetoric of a global War on Terror was used to justify the intensification of surveillance activities domestically as part of the global conflict. In other words, I focus less on the specifics of the military operations associated with the War on Terror and more on the justification for the expansion of surveillance, justification meant to influence the public imagination about the role of surveillance within Canadian domestic life. As Lyon writes of the escalation of surveillance after 2001, "we can safely note that prominent among the technical responses to 9/11 is the reinforcement of surveillance. With better systems in place, their promoters insist, the risk of another attack could be drastically reduced. With better means of identifying, classifying, profiling, assessing, and tracking the population, the chances of preventing future attacks may be increased."[44] The idea of preventing future terror attacks became the political, military, and public mantra of the rhetoric of the War on Terror. Yet the conception of terror remained ill-defined: terror – as the target of surveillance operations – was a nebulous and moving spectre.

The notion that terror was anywhere and everywhere became a shorthand to justify state policies and military actions, as well as to vilify individuals and communities through orientalizing narratives of terror, which were anchored in racism and Islamophobia. The rhetoric of the War on Terror became part of the public imaginary and political landscape in Canada that was used to rationalize a diverse range of laws, policies, procedures, and actions in the name of national safety and security against an ill-defined "terror." This is not to say that this rhetoric remains in the abstract; surveillance legislation, policies, and activities of the state and its corporate arms have had material and real consequences, particularly for those who are Black, Muslim, Indigenous, racialized diasporic, 2SLGBTQIA+, and disabled. Indeed, I submit that the artworks in this book collectively reveal that the War on Terror is Canadian settler colonialism by another name. These artworks address contemporary surveillance as a vehicle for settler colonialism, and in its overlap with other systems of oppression internationally and, specific to my focus in this book, in Canada. Following sociologist Sean Hier and communications studies scholar Josh Greenberg, I look to art as it reveals "how surveillance practices are related to political and social problems in the context of material relations of power and institutionalization."[45]

Chapter Organization and Art's Contribution

States of Observance brings together a wide variety of creative projects with different styles, methods, forms, and subject matter to examine the developing body of contemporary surveillance art in Canada. The basic premise of this book's methodology stems from two questions: What does art do? And, what conversations about surveillance and its logic does art engage in? Considering art and creative practices in these terms asserts their capacity to produce meanings about, rather than simply respond to, socio-political issues. In many ways, *States of Observance* arose from an interest in the capacity of art practices to generate and provoke new ways of understanding the structures of surveillance in the context of Canada's participation in the security conventions of the War on Terror. The book was also driven by a curiosity about the kinds of knowledge that might emerge from placing these creative works in direct conversation with surveillance studies research and further contextualizing them through insights from an interdisciplinary group of authors as well as media content and government policies. To bring these sources together, I ask three main questions:

1. What are the different approaches and themes that recur within Canadian artworks and exhibitions, and what do such approaches reveal about the politics of surveillance?
2. In what ways can the themes and ideas brought forward by art and creative practices help shape and reshape understandings of surveillance within contemporary Canada?
3. In light of knowledges being produced by creative works, how might we rethink Canada's historical and ongoing investments in surveillance as part of the creation and maintenance of the nation as a settler colonial state?

I thread these questions through the different chapters, subthemes, and analyses in an effort to bring creative practice, socio-historical context, and national histories together to register art's contributions to understanding surveillance in Canada from 2001 to 2021.

A major goal of this book is to introduce readers to a wide range of critical creative projects. While it is not intended to provide a comprehensive listing of all creative works that deal with contemporary surveillance in Canada, this book seeks to provide a broad catalogue of many of the works addressing this topic, bringing them together to create and discover meanings in their juxtaposition. My approach borrows much from critical curatorial practice and exhibitionary methodologies, in that I see the process of reading the works against one another as a form of critical inquiry.[46] I am inspired in this organization by exhibitions that bring together a variety of creative projects under a thematic umbrella. Within Canada, multiple group exhibitions have brought to life the art of surveillance to address the complexities and nuances of surveillance as a point of critical analysis. These include *Sorting Daemons: Art, Surveillance Regimes, and Social Control* (Kingston, 2010; curated by Jan Allen and Sarah EK Smith); *Cyber-Surveillance in Everyday Life* (Toronto, 2011; curated by InterAccess Gallery); *Border Culture: Part Three (Security, Surveillance)* (Windsor, ON, 2015; curated by Srimoyee Mitra); *Surveillance* (Calgary, 2015; curated by Su-Ying Strang); *Security in our society: what remains of our personal freedoms?* (Montreal, 2015; curated by Art Souterrain); *Scaffolds I can no longer see* (Toronto, 2019; curated by Sophia Oppel and Philip Leonard Ocampo). The curators of these shows, among others, bring together works by different artists who utilize multiple strategies, contexts, forms, and mediums to think through the contours of surveillance. In juxtaposing artworks in new ways, these exhibitions are able to offer new insights and analysis. Similarly, in this book, I use the intent and interpretation of each artwork to reveal different themes of surveillance and its logic, and then place these artworks in conversation with each other and alongside scholarly and public debates. A curated collection of artworks produces knowledge that extends beyond the sum of its parts. Here, it introduces readers to the larger body of artistic works being produced on contemporary surveillance, while also situating these works within broader fields of scholarly research and socio-political contexts.

To interrogate how art sheds light on surveillance logic in a context of settler colonialism and the War on Terror, I rely on an interdisciplinary group of scholars, writers, and artists – many of whom are operating from positions and subjectivities to which I cannot speak – to bring forward complexities, experiences, and analyses. To do this, I utilize a citational ethics of relation, as French literature scholar Annabel Kim calls for, rather than one of erasure or property.[47] This strategy is based on questioning how histories and narratives are told, and by whom. Throughout, I draw on interventions from a number of fields – such as Indigenous studies, critical Muslim studies, queer studies, Black studies, disability studies, gender studies – that are examining specific contexts that can be distinct from my own research focus. These works present frameworks and critiques that are crucial to understanding the different conversations and analyses of the artworks and the structures they address in relation to the logics of surveillance. I draw on this scholarship not to remove the work from its original context but to engage and acknowledge these frameworks as critical strategies for illuminating the intersecting contexts of art, surveillance, and the settler colonial project of Canada.

I approach this strategy with caution, aware of the histories of excavation and appropriation of the white academy that claims ideas, voices, and stories as its own. Here, I am cognizant of what Indigenous studies scholar Eve Tuck and critical pedagogies and decolonization scholar Wayne Yang identify as "settler moves to innocence," a way in which white settlers attempt to absolve themselves of colonial complicity.[48] Art historian Erin Morton emphasizes white settler scholars' attempt to "move to settler innocence ... to critique national settler texts," and highlights how strategies "must always be approached with a clear intent."[49] My intent here is to acknowledge my own complicity within the settler colonial project, and to dialogue with anti-racist, Indigenous, decolonial, 2SLGBTQIA+, and disabled writers and scholars to interrogate art and the "national settler texts" of state and surveillance. I work in conversation with these authors to enrich and build a decolonial critique of the settler colonial project in Canada and its use of surveillance logic as impermanent, and to reveal its ongoing fragility, construction, and maintenance, challenging its assumed stability.

Each of the four chapters of this book examines a particular shared point of address brought forward by the artworks. The first three – "Gazing," "Policing," and "Landing" – unite the creative projects through the ways in which they incorporate certain representative elements of surveillance systems that I put in conversation with the dominant belief systems of settler colonialism in Canada. In chapter 1, "Gazing," the frame brings together artworks that take on the surveillance gaze and structures of watching and being watched as modes of control. The surveillance gaze, however, is not just about the capacity to see – or the act of looking at – what is visible; rather, it is tethered to histories of who gets the authority to look, to recognize, to claim, and to police, histories tied to the dominance of white settlerhood and colonial imaginaries in the Canadian context. Chapter 2, "Policing," highlights artworks in relation to what I term "surveillant policing." It emphasizes the interconnections between surveillance and policing activities, which are mutually motivated by the perseverance of state (and non-state) authority and by upholding settler colonial visions of the Canadian nation-state. Chapter 3, "Landing," examines artworks that connect surveillance to ongoing histories of dispossession of Indigenous life and to impositions of white settlerhood within Canada. Specifically, the idea of landing in artworks questions how settler colonialism in Canada is produced and reproduced through surveillance visualizations of the land. While the central idea in the first three chapters is developed around the subject matter of the artworks, chapter 4 – "Resisting" – is organized somewhat differently. Instead of featuring a representative subject, this chapter brings works together through a shared strategy – a strategy to counter, resist, and create alternatives to hegemonic surveillance systems.

Many of the creative projects could appear in multiple and even overlapping chapters, but I have placed them within certain chapters to draw out what I see as complementary conversations with other works. Within the chapters themselves, I break down the larger theme into smaller and more focused units, again to analyze different artworks through the complexities they bring to the study of surveillance, and the unique vision they bring in their juxtaposition to each other. The organization of the chapters and the artworks within them is not meant to constrict meanings or foreclose other conversations; rather, it identifies what I see as recurring trends in art that address the Canadian surveillance context. Much like a curatorial strategy deployed within an exhibition space, this structure brings different artworks together around a particular topic to discover what meanings and knowledge they synthesize and create.

The thirty-three artworks in this book represent a variety of perspectives, positionalities, and approaches to surveillance systems in Canada as part of the War on Terror and ongoing colonialisms within the nation-state. While each deploys different strategies and mediums in its creative and critical explorations of surveillance logics, they come together in conversation about how surveillance systems have come to shape and reshape Canadian settler colonial history in a contemporary moment defined by the War on Terror. They reflect, I argue, that this period both heralds new ways of deploying surveillance strategies and relies on a historical and ongoing logic that has informed the role of the state in Canada since settler occupation. Certainly, they demonstrate the generative capacities of art and creative practice as a methodology, in that art opens up rather than closes off discussions and politics. Art is a critical intervention that presents pathways to trouble, reveal, challenge, and resist surveillance logic and its interconnections with settler colonialism as foundational to the Canadian nation-state.

Gazing

1

One of the most powerful strategies of imperial dominance is that of surveillance, or observation: because it implies a viewer with an elevated vantage point, it suggests the power to process and understand that which is seen, and it objectifies, and interpellates, the colonized subject in a way that fixes its identity in relation to the surveyor.

Bill Ashcroft, *Post-Colonial Transformation*

This chapter focuses on artworks that foreground forms of surveillance seeing and the ways in which the logic and aesthetics of surveillance produce meaning. It puts artworks that explore the characteristics and structures of surveillance in conversation with scholarship that examines how the settler colonial state produces and reproduces itself, in order to reveal how deeply entwined state violence, epistemology, and technology are in systems of domination and oppression. Specifically, this chapter examines contemporary creative projects that centre the "surveillance gaze" as essential to engaging with structures of Canadian settler colonialism. The nine works in this chapter come together under the umbrella of the surveillance gaze in that they each examine how the rhetoric and aesthetics of surveillance perspectives are understood. They critique the processes by which the surveillance gaze – in its monitoring of public and private spaces – produces, identifies, and legitimates good citizens while controlling and excluding those who do not fit within the bodies, behaviours, and expectations of social, cultural, and national belonging in Canada. In this way, the artworks reveal that the "gaze" in surveillance is not just visual; it is systemic.

The concept of a "surveillance gaze" has been a key point of analysis in the field of surveillance studies.[1] This concept builds on studies on "the gaze" – which visual culture scholars Marita Sturken and Lisa Cartwright define as the ways in which a viewer "is situated in a field of meaning production (organized around looking practices) that involves recognizing oneself as a member of that world of meaning" – and brings it into the context of surveillance logic.[2] The surveillance gaze is not just about the capacity to see or the act of looking at what is visible; rather, it is a form of visualizing or understanding the world that is inextricably linked to structures of power that narrate history in specific ways. The surveillance gaze is about power and control and is integral to a larger system of surveillance visuality, which refers to how social conditions are constituted through visual discourse and become visually knowable. The surveillance gaze references the angles of watching and being watched that are attached to controlling, monitoring, policing, and, ultimately, to knowing. Within the context of the settler colonial project of Canada, the surveillance gaze operates as a mechanism of power that enables dispossession, enforces hierarchies, and sustains the settler state. It is the gaze that binds surveillance logic to settler colonial perspectives and interpretations, and makes apparent that surveillance viewing is settler colonial viewing. Many of the artworks in this chapter use a surveillance aesthetic that is visual, and I use language around seeing, viewing, and watching to discuss them. I want to emphasize, however, that, as part of recognizable surveillance aesthetics, these visual aspects are tightly bound to other senses and knowledge systems that together form a process of generating access, control, and understanding.

The term "surveillance gaze" has received an enormous amount of scholarly and public attention since 2001, as surveillance systems have increased and expanded their reach beyond militarized contexts into everyday behaviours, spaces of consumption, and civilian life. In the cultural and political aftermath of 9/11, the surveillance gaze became more pronounced in the public space, expanding to subsume a broader and more refined sense of what people and activities needed to be monitored for the public and national good. The surveillance gaze as an apparatus of authority and control, of course, predates 2001. Surveillance seeing is not new. While this contemporary expansion is linked to and affirmed by the rhetoric and policies of the War on Terror – and thus many studies (including this one) focus on this particular historical context – it is important to recognize that surveillance as a mode of viewing emerged much earlier than the early twenty-first century. Many works coming out of the field of surveillance studies examine how surveillance technologies use visibility as a control technique in the modern era, often applying French philosopher Michel Foucault's articulation of the panopticon as a social theoretical model.[3] Based on architect Jeremy Bentham's eighteenth-century architectural design of a prison structure that enables all prisoners to be observed at all times by a single, hidden guard while the prisoners never know if they are being seen at any given moment, Foucault's panopticon has become a cultural symbol for understanding the disciplining effect surveillance has on modern society. Foucault offers the panopticon as a way to think about how looking and visuality are key participants in modern structures of authority.[4] Indeed, sociologist Sean Hier defines "the desire to coordinate and control populations" that informs the modern surveillance gaze as "the panoptic impulse," and surveillance studies scholar David Lyon argues that "the disciplinary power of the Panopticon is the archetypical power of modernity."[5]

Other scholars in surveillance studies take the concept of the panopticon, which is still a top-down, centralized surveillance vision, and push it further by positioning surveillance monitoring as part of a multidirectional, decentralized conception of surveillance looking, often conceived as the polyopticon.[6] In their respective writings, sociologist David Wood and criminologist Majid Yar call for a critical reinterpretation of panopticism that connects aspects of looking, being seen, and visualities to other theories of control, such as bio-politics, and to critiques of institutional structures.[7] In her book *Dark Matters*, African and African diaspora studies and surveillance studies scholar Simone Browne asserts the necessity to locate panopticism as "a social practice," one that accounts for systems of surveillance and control in relation to broader social categories of race, gender, class, nation, and imperial domains.[8]

I position the artworks in this book in conversation with some of these critical reinterpretations of panopticism that seek to connect visual aspects of control to other modes of understanding the operations and policies of surveillance systems, particularly in relation to Canada as a settler colonial project and the War on Terror. I include in this chapter a series of artworks that centre the surveillance gaze so that they may address surveillance in the contemporary Canadian context within the socio-historical reconstitution of what art historian Kirsty Robertson identifies as "vision in relation to power and resistance."[9] Here, the creative projects use the gaze to show how the visualities of surveillance are inseparable from interlocking regimes of authority and control that organize the disciplinary power of the settler colonial state.

The surveillance gaze operates in relation to other structures of power and oppression and creates visual codes of monitoring, interpreting, containing, and policing that affirm and justify them. In so doing, the surveillance gaze intersects with other gazes, such as the colonial gaze, the male gaze, the white gaze, the classed gaze, the cisheteropatriarchal gaze, and the ableist gaze. Certainly, these different forms of "the gaze," which claim an authoritative bio-politics, inform and bolster one another and justify cultural, political, and structural systems. In Canada, as gender and social justice scholars Carla Rice, Susan Dion, Hannah Fowlie, and Ingrid Mündel succinctly describe in their essay addressing intersections of the heteropatriarchal settler colonial gaze, "colonial processes are deeply intertwined with heteropatriarchal ones; through state mechanisms ... such as the Indian Act and residential schooling, settlers imposed heteropatriarchal structures – the nuclear family, the heterosexual matrix, the masculinist political economy."[10] The surveillance logic within the settler colonial context of Canada reveals that "techniques of observation, discipline, and control are deeply rooted in the very formation of the state," as white settlers claim the right to look and thus determine inclusion within national boundaries.[11] As literary scholar Michael R. Griffiths asserts, it is "white settler subjects [who] claim the authority to recognize or disavow."[12] Ultimately, the surveillance gaze demarcates forms of belonging and inclusion and, in Canada, reveals what scholar of politics and government Ahmad Sa'di describes as the "colonial desire for panoptic control."[13] In this sense, while the overall concept of the gaze refers to the power to look, name, claim, and control, the specific form it takes – surveillance, colonial, white, cisheteropatriarchal, ableist – determines the emphasis, or what information is gleaned from the process of looking. Significantly, this emphasis does not preclude those systems from overlapping and informing one another. When I use the term "surveillance gaze," I refer to the ways in which surveillance and its forms of looking operationalize settler colonial structures and their intersecting systems of white supremacy, racism, capitalism, and cisheteropatriarchy. While the artworks in this chapter do not necessarily address each form of oppression directly, I argue that the intersections they do uncover work to implicate mutually reinforcing, exclusionary systems of power.

The artworks in this chapter critically foreground representations of surveillance visualities, signalling the surveillance gaze through emblematic characteristics – such as top-down vantage points or mundane video feeds – that readily convey contemporary surveillance to a broad audience. The surveillance gaze they reveal, however, is not merely an aesthetic or neutral act of observation; it is a mechanism through which power is exercised, disciplining bodies and shaping behaviour according to normative frameworks. Media studies scholar Lisa Nakamura highlights that, in its process of "classifying some bodies as normative and legal, and some as illegal and out of bounds," surveillance "*remakes* the body as a social actor."[14] In other words, as visual communications scholar Jonathan Finn explains, "surveillance is not simply a technology or apparatus employed in state-sanctioned programmes of social control." Rather, he argues, "it has become a social practice: a way of seeing, understanding and engaging with the world around us."[15] Surveillance has become a pervasive and normalized aspect of everyday life such that we move through public spaces with some ambient awareness of being watched and in turn unconsciously or consciously assess ourselves and one another according to internalized surveillance logic derived from the Canadian settler colonial project's norms of acceptability. Finn conceptualizes the connection between the normalized, societally embedded surveillance gaze and structures of power and control as "seeing surveillantly."[16] Building on this idea, I see the concept of the surveillance gaze as a discourse; the concept refers to the larger discourses of power and control that are tied to forms of seeing and being seen through surveillance structures, and it represents a mechanism through which social relations and systems of oppression bring to life certain ways of being in the world.

The creative practices included in this chapter draw on the social and historical familiarity of what it looks like to see via surveillance technologies – that is, via distinct yet related modes of viewing that have become recognizable, familiar, and authoritative in the contemporary moment. In most cases, the creative projects enact a consciously inexact copy of these visual codes as a way to play with the liminal state between knowing and unknowing, familiarity and unfamiliarity, expected and uncertain. This artistic strategy distorts recognizable and societally ingrained forms of the surveillance gaze and encourages viewers to, first, notice them, and second, question the systems that normalize them. By distorting the angles of vision tied to unexamined expectations of surveillance, the artworks lay bare the role of surveillance logic in the processes of power that uphold the settler colonial project of Canada during the War on Terror.

This chapter is organized into three sections, which bring into focus different dimensions and approaches to the surveillance gaze: security, seeing, and being seen. In "Security," through examining the role of surveillance in creating or destabilizing a sense of safety and well-being, the artworks reveal how the surveillance gaze governs interpretation and the power inherent therein. In "Seeing," the artworks reintroduce embodiment to the seeming objectivity of the surveillance gaze, considering the bodies of those who watch and the subjective human experience of seeing surveillantly. In "Being Seen," the creative projects implicate the audience within the piece as both subjects and objects of the gaze to question authoritative looking and denormalize our complicity in it as the one who is watched. Although the nine artworks are divided into these three sections, they speak across thematic divisions and bring forward larger ideas that complicate and critique the surveillance gaze, as well as mobilize it as a form of critical address of the settler colonial context. They offer unique insights into the inextricable relationships between surveillance colonialism, white supremacy, capitalism, and cisheteropatriarchy, and illuminate how surveillance logic and its gaze participate in shaping and reshaping settler colonialism in Canada.

Security

The artworks in this section take the visual conventions of surveillance and distort them to question for whom and under what conditions surveillance looking is applied as a form of security. The four works use surveillance watching itself as a point of entry, which I place in relation to the angles of vision associated with surveillance structures in the settler colonial context. In Canada, the surveillance gaze operates as a technology of securitization that maintains spatial and racial boundaries, working in tandem with the colonial state to reassert its authority and ensure the continued dominance of white settler populations. Surveillance logic is deployed not only to observe but also to discipline and dispossess. What the artworks here question is the promise of security. They ask: What is the presumed role of surveillance in securing safety? Who is made secure by the surveillance gaze and its apparatus within the settler colonial Canadian context? And in what ways does invoking security as a rationale for surveillance give rise to a surveillance gaze that is self-reinforcing?

Each of these works materialize Finn's "seeing surveillantly" in that they position the audience member as the one who sees – the one who deploys the surveillance gaze – and they trouble the assumptions underlying this form of information gathering. The works highlight that what is revealed by looking through surveillance vision relies heavily on the "eye of the beholder" – the interpretative and subjective context through which images are understood. They play on that position by compelling audiences, as scholar and activist bell hooks argues in relation to Black audiences considering representations of Blackness on film, to "consider the perspective from which we look."[17] What assumptions does the surveillance watcher bring to the watching, and who gets to be the one who wields such power? The surveillance gaze here is deeply entangled with the logics of security that police the settler colonial context and assess threats to the settler state's territorial and political authority. Ultimately, these works challenge the way the gaze is framed as ensuring security to justify surveillance as a required mode of inquiry. In this, I argue that these artworks, either directly or by implication, also interrogate settler colonialism and its mutually constitutive systems of oppression and inequity as interpretative contexts that underscore meanings produced from surveillance images in Canada.

Toronto-based painter Michael Lewis produced an oil-on-canvas work that showcases a familiar angle of vision tied to surveillance technologies. His painting *Some will take more prodding. Others will be more difficult* (2008, fig. 1.1) shows how easily representative characteristics of the surveillance gaze can communicate suspicion and criminality, and it demonstrates the level of control the surveillance viewer has in interpreting the images. Lewis's painting depicts a trade show environment populated by gatherings of people talking and milling about. The vantage point of the composition is angled downward, reminiscent of common forms of security or CCTV cameras that monitor public and private spaces. The scene includes a series of trade show booths – their large advertising banners rendered unreadable in their depiction – located in a high-ceilinged auditorium or meeting hall, signifying capitalist productivity. Groups of people mill, gather, and sit among the booths, talking and gesturing in subdued, appropriate, and professionalized ways. Everyone included in the scene is dressed in the monochromatic and indistinguishable corporate attire of grey and black suits and dresses. This subdued colour palette extends beyond the dress of the figures into the painting itself, lending the scene a cold cast often associated with the black-and-white technological gaze. The representation of such a familiar and nondescript activity provides a sense of the mundane and the everyday, and raises questions as to why such a scene would need to be surveilled at all. Yet the

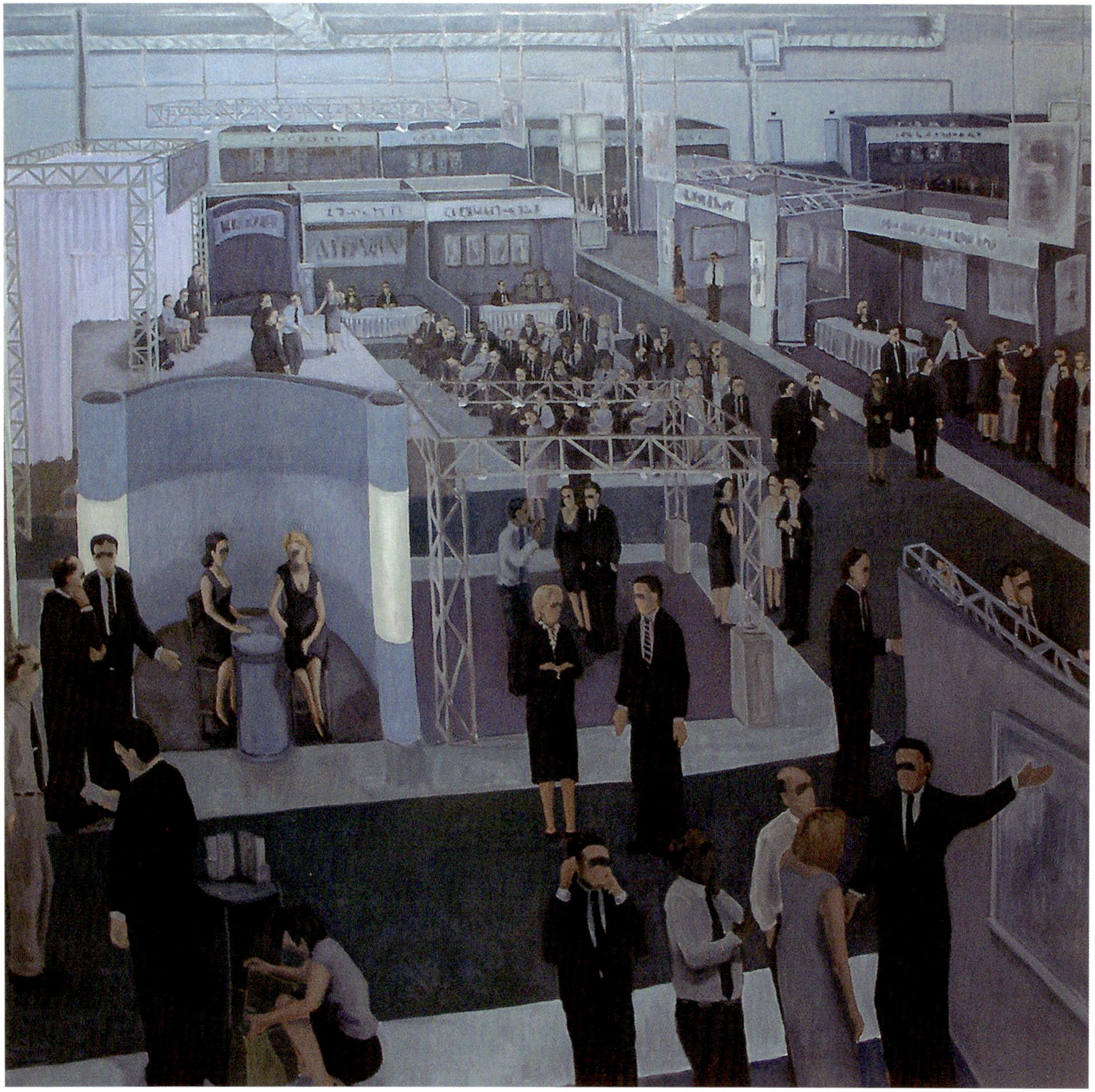

1.1 • Michael Lewis, *Some will take more prodding. Others will be more difficult*, 2008.

angle of vision looking into the scene frames the activities as in need of surveilling and, therefore, as potentially sinister.

The eyes of all the figures in the scene are obscured by black squares, invoking visual codes of censorship traditionally used to conceal or protect identities. The blacked-out eyes also alter viewing access into the faces of the figures, rendering them visually similar to one another. Describing the painting in the catalogue for the group exhibition *Sorting Daemons: Art, Surveillance, and Regimes of Social Control*, held at the Agnes Etherington Arts Centre in Kingston, Ontario, in 2009, curator Jan Allen argues that the blacked-out eyes of the subjects point to the suppression of "their individuality, not [the protection of] their privacy"; Allen sees the black squares as "symptoms of an absence of will."[18] By covering the eyes of all the figures in this repetitive way, the artist removes individual identity from the piece. For Allen, the surveillance gaze operates not as a means of protecting and ensuring the safety of the individuals in this space, but rather as a method to track and regulate their bodies and behaviours in order to maintain an easily controllable status quo.

The surveillance gaze here is not merely about watching; it is about categorizing, regulating, and rendering lives as legible to the settler colonial and capitalist state while stripping them of autonomy and subjectivity. This abstraction of identity reflects how surveillance reduces individuals to data points, part of a larger settler system that classifies who is included in and excluded from the nation-state, and under what conditions. Importantly, the visual detail of the blacked-out eyes in the painting not only hides identity but also introduces the idea of criminality. The censorship of some details of the piece, namely the eyes of all the figures, provokes viewer questions as to what is illicit or forbidden that requires removal and why. In this way, *Some will take more prodding. Others will be more difficult* illuminates the failed experiment of surveillance: rather than evoking feelings of security, its own visual codes introduce feelings of insecurity and uncertainty. The visual detail of the blacked-out eyes makes the figures look suspicious, imposing the idea of criminality onto individuals. Lewis applies recognizable visual markers of surveillance – the surveillance angle of viewing and figures marked as suspicious and warranting surveillance – to repurpose the mundane corporate environment as potentially insidious and criminal. He uses aspects of the surveillance gaze to show how easy it is to associate such viewings as criminal: it is not necessarily the bodies, actions, or environments that encode this scene as socially unacceptable, but rather it is the lens through which this scene is viewed that adds a tinge of uncertainty, suspicion, or curiosity.

At the same time that the visual characteristics associated with surveillance systems introduce suspicion and criminality to those caught in the lens of the surveillance gaze, they also demonstrate the power and authority that the gaze holds. The power of the watcher in interpreting surveillance images lies in their authoritative role in assigning meaning, value, and intent to what is seen. The gaze of this piece travels in one direction: from an omnipresent viewer toward the figures in the scene. These figures do not look back at the surveilling viewer, so the viewer is rendered invisible and unseen to them. In this, the painting emphasizes the potential power and control of the surveillance gaze to operate without judgment or reproach. Hidden behind the undetected eye of the surveillance apparatus, the viewer is able to deploy this gaze without being seen or looked back at; this viewing dynamic illuminates the uneven power relationship behind seeing surveillantly, where the viewer holds agency over the viewed in forming narratives and meaning.

Some will take more prodding. Others will be more difficult reveals how the surveillance gaze functions not merely through the passive recording of images but also through active processes of interpretation by those in control – state agents, security personnel, or institutional authorities – who, of course, determine what constitutes suspicious behaviour, criminality, or threat. This interpretive power allows watchers to construct narratives that may go on to become justification for interventions, exclusions, or violence. In the context of Canadian colonial-capitalist systems, such interpretation, then, materially reinforces asymmetrical power dynamics inherent in the surveillance logic of settler colonialism, with its attendant systems of white supremacy, capitalism, ableism, and cisheteropatriarchy. The watcher thus holds a dual authority: one over the physical space and bodies being monitored, and another over the discursive space in which those bodies are understood and acted upon. This dual authority highlights how the surveillance gaze operates not only through visibility but also through the interpretive frameworks that give that visibility political and material consequences. This artwork places its audience in a position to think about questions of whose bodies belong, what rules structure ideas of acceptability, who gets to monitor whom, and how the surveillance gaze itself interpellates those captured by its lens as potentially sinister or criminal and therefore in need of clandestine monitoring to ensure public safety and security. Importantly, it is not the figures within the scene of the image that benefit from this promise of security.

Wanda Nanibush's piece *Carrying* (2010–16, fig. 1.2) also plays with conventional surveillance angles to interrogate the systems of control and authority associated with the power to look, and questions the promises of security of these systems. Nanibush mounted this piece at the SFU Gallery at Simon Fraser University in Burnaby, British Columbia, in 2016 as part of the group show *Unsettled Sites*, curated by Tarah Hogue. The larger exhibition examined, in Hogue's words, "the complex entanglements of belonging and refusal from both settler and Indigenous perspectives. The works disturb the ubiquity of settler colonialism, the violent cleaving of Indigenous inhabitants from the land through both physical removal and forced cultural assimilation."[19] *Carrying* includes a live video feed of the artist's home territory of Chimnissing, the Beausoleil First Nation (Anishinaabe) community in the southern Georgian Bay region of Ontario. The video installation displays footage from the security camera mounted on top of the band council's office in Nanibush's community. Bodily movements in the gallery trigger the video's sound, an artistic choice that makes clear, material connections between those who view and how the viewing takes shape. Audience members in the gallery space can watch the activities and people on a ferry dock in the Beausoleil First Nation community in real time. The projected video image is immediately recognizable as generated by a security camera, which is at a distance and angled downward. *Carrying* provides a single vision: its view of the ferry dock narrows the audience's focus, requiring them to make meaning of this particular location and its activity. Within the gallery space, the angle

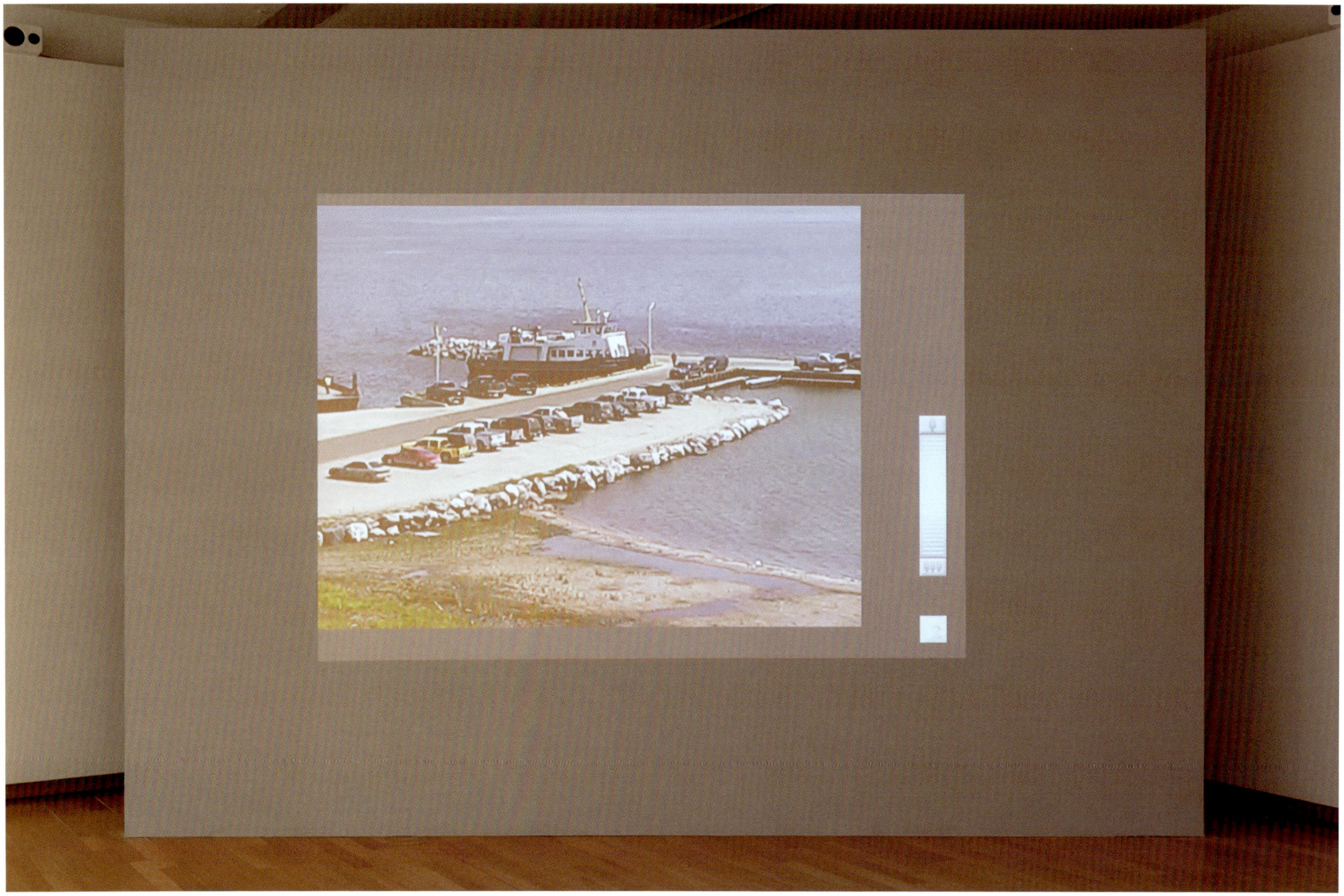

1.2 • Wanda Nanibush, *Carrying*, 2010–16.

of vision and access to the scene illuminate the uneasy juxtaposition between, as Hogue articulates, the artwork's "function as an integral part of the daily operations of the band versus … our desire to consume it as an art object or artwork."[20]

This juxtaposition of real-life security and artworld contexts is reinforced by the fact that live footage from multiple cameras operated by the band council, including the one used in *Carrying*, is publicly available on the Beausoleil First Nation website.[21] Given that these images are readily accessible via the internet, what, then, does a work like Nanibush's offer when it reproduces some of them in an art gallery? Placed in the context of a gallery space, these surveillance images take on the role of an artwork, prompting the act of looking but also of contemplation. As artist Raymond Boisjoly and curator Jonah Gray write in their review of the exhibition, the focused surveillance angle emphasizes the "mundane movements of the people" in the image and gives these activities "heightened importance when viewed in the art-gallery context; viewers are trying to understand what makes this meaningful. These peoples' need to simply move from one place to another … takes on this complexity."[22] This artwork's reception is further complicated by its relationship to the other people and objects present in the gallery space. People directly accessing surveillance images from the band council's footage on the internet would view these images privately on a small personal device. In contrast, gallery visitors engage with the footage publicly, on a large scale, among other gallery visitors. *Carrying*, then, can be read against the larger theme of the exhibition, which, in the curator's narrative, is related to "haunting." Such a theme leads us to ask, what legacies and histories haunt the ways in which audiences interpret images? Within the settler colonial context of Canada, these haunting histories are inextricably linked to surveillance tactics that monitor, dispossess, assimilate, and eliminate Indigenous nations and communities from the land as part of securing the settler nation-state, but they are also related to Indigenous people refusing, resisting, and surviving these tactics. Thus, the exhibition theme shapes the viewing of these surveillance images in a way that complicates an easy connection between surveillance and security.

Nanibush's *Carrying* emphasizes the contemporary life of surveillance images that are used within a community by that community, in this instance by the Beausoleil First Nation. The artwork illuminates the ways in which the First Nation itself utilizes surveillance systems to secure its own members and community. It is important to note that, while the visual codes of the surveillance gaze may be similar, the logic of this system is politically and ethically different from the surveillance logic of settler colonialism. While the Canadian settler state employs surveillance structures to monitor, manage, and regulate Indigenous presence and resistance, thereby facilitating the ongoing project of territorial occupation and settler authority, here Indigenous surveillance systems assert sovereignty, safeguard land and people, and revitalize self-governance structures that have been undermined by colonial intervention. Certainly, the visual codes of the surveillance gaze in *Carrying* resist and reframe surveillance systems within an Indigenous context at the same time as they gesture toward longer and ongoing histories of Canada's violent use of surveillance against Indigenous people.[23]

Carrying, then, brings into focus a tension between those insidious and violent histories of surveillance targeting Indigenous communities and the decision of an Anishinaabe community to self-surveil. It raises questions about who has access to looking, who is being looked at, and whose security and safety are ensured by that looking. In Nanibush's piece itself, the question of the relationship between the viewer and viewed is pronounced, highlighting the distinction between the surveillance viewer and the surveillantly viewed. Those captured in the video feed in *Carrying* are unaware of their role in the artwork (and perhaps in the security capture more generally) and are therefore non-participants in the work. The gallery audience, on the other hand, are active participants in the piece, their very movements activating the sound for full viewing of the work. This power dynamic of viewer versus viewed is common within surveillance structures, but *Carrying* pushes it further by setting up surveillance access to the people and activities of an Indigenous community in Ontario from the distance of an art gallery in British Columbia. And, as Boisjoly and Gray point out, the viewing dynamic changes based on whether the audience members and people depicted in the piece are Indigenous or non-Indigenous.[24] In this way, *Carrying* confronts viewers with their own unique relationship and proximity to promises of security within the settler colonial context of Canada.

While Lewis's and Nanibush's works use overt angles of the surveillance gaze to question the complicated connections between surveillance and security, artist Antonia Hirsch rethinks the very qualities of seeing at the heart of the presumed gaze of surveillance structures. Her series of two projects she produced in 2008 and 2009 repurposes domed security mirrors – the type that are often placed in the corners of hallways or staircases to provide additional angles of visibility. In her piece *Double Blind* (2008, fig. 1.3), Hirsch mounted these mirrors on the side of a high wall at Vancouver Community College (VCC). Her artwork *The Invisible Hand (after Adam Smith)* (2009, fig. 1.4) similarly installs the mirrors but within gallery settings, including in *Sorting Daemons*, the exhibition in which Lewis's *Some will take more prodding. Others will be more difficult* was also exhibited. For each of these works, Hirsch arranges the domed mirrors on vertical surfaces to create braille patterns of different sizes in imitation of the Snellen chart that optometrists use to test visual acuity. As Hirsch herself articulates, the works present a conundrum of access and perception: those who can see the mirrors are unlikely to read braille, while those who rely on touch to read braille are physically unable to reach or decipher the oversized mirrors.[25] By requiring viewers to both see the image and visually interpret braille, Hirsch exposes the assumptions and inequities built into vision-centric surveillance systems. The works critique the presumed universality of sight as a mode of knowledge and authority, showing how tools designed to expand visibility – domed mirrors – produce selective access and reinforce structural limitations. In doing so, Hirsch reveals the inaccessibility and exclusions embedded in systems of surveillance and security, challenging the idea that increased visibility equates with understanding or control.

While both of these works use the same materials and form, their content differs. *Double Blind*, as writer Vanessa Kwan describes, "reveals the inaccessibility of mainstream communication for the blind and relates to VCC's programming for visually

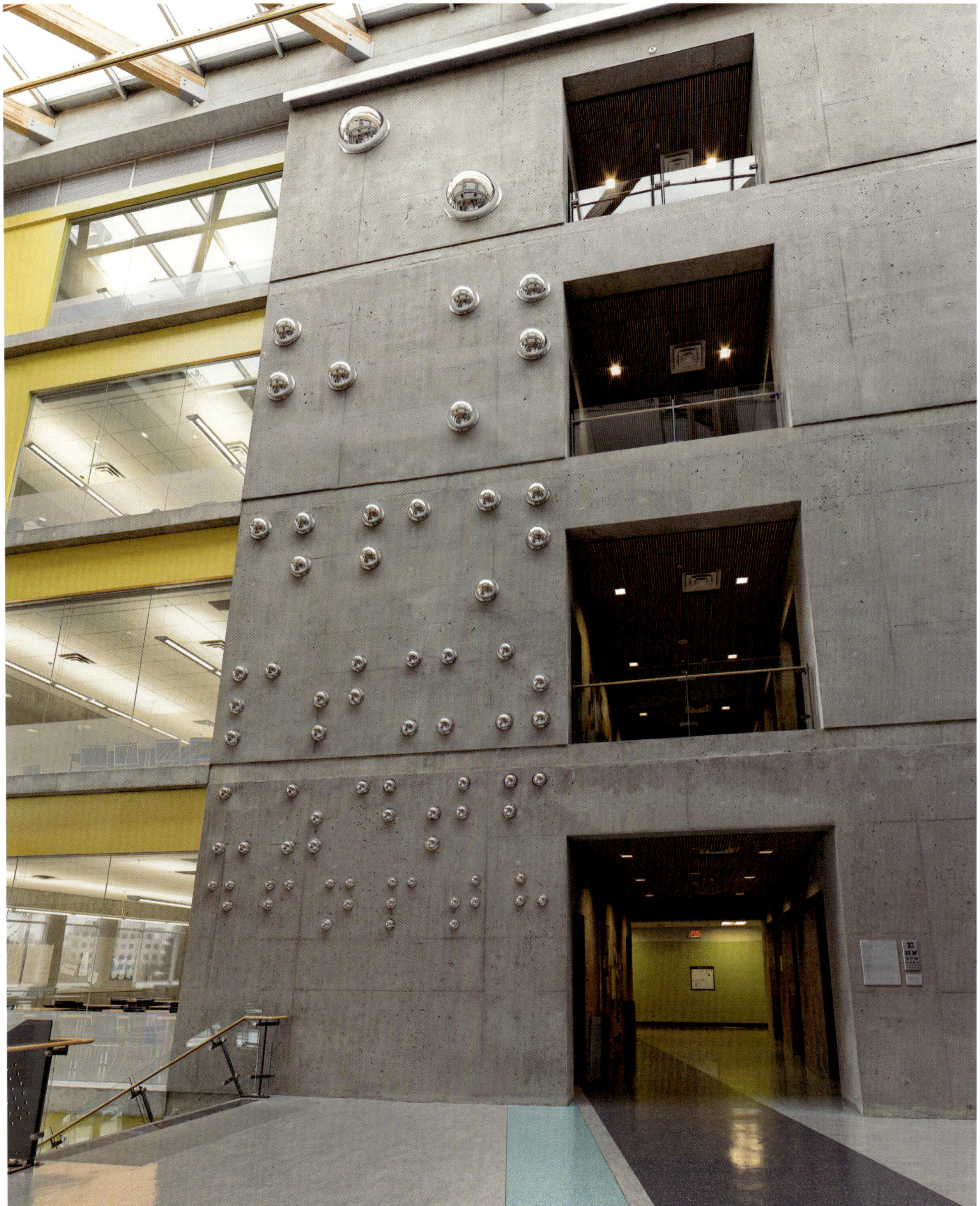

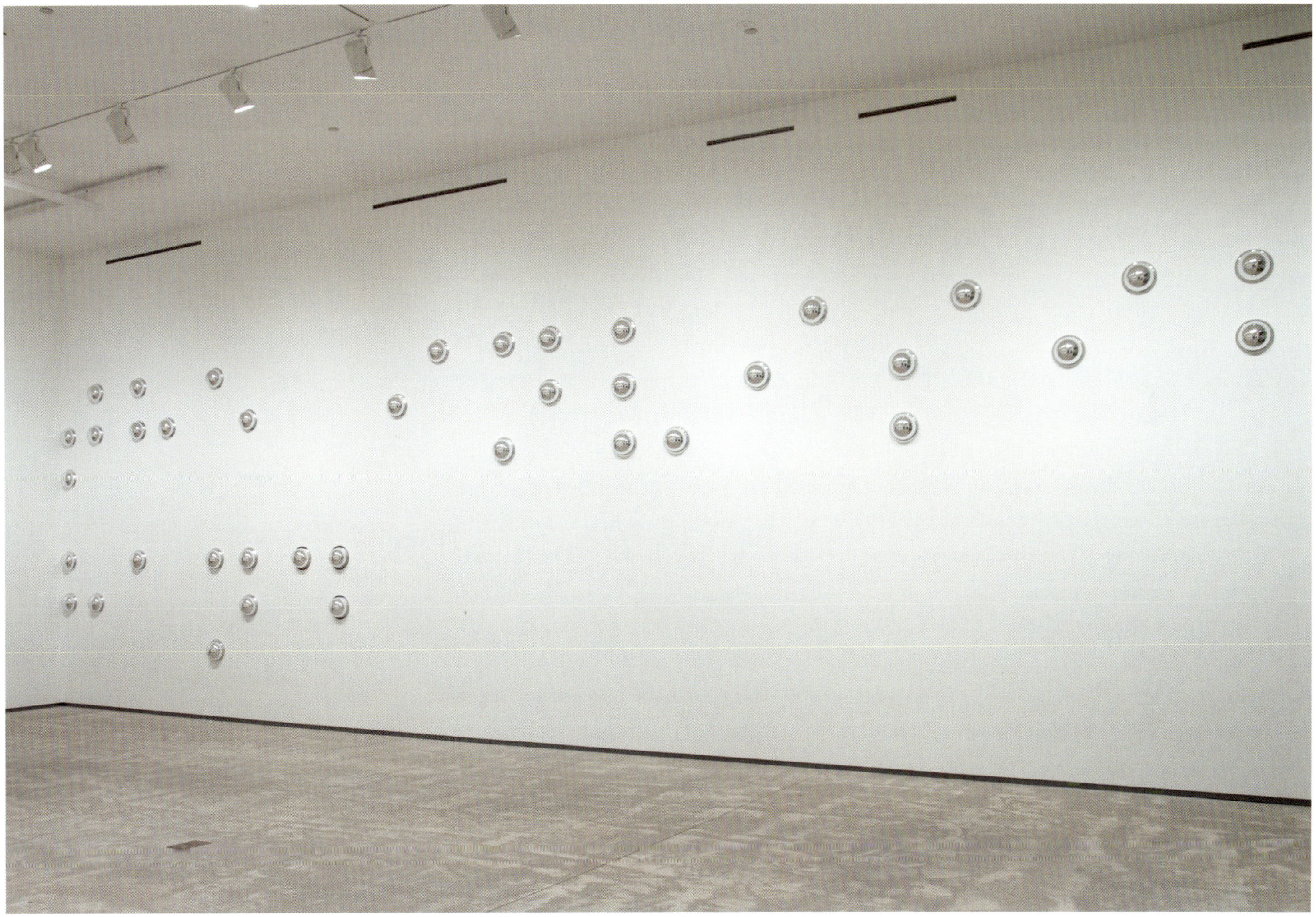

opposite **1.3** • Antonia Hirsch, *Double Blind*, 2008.
above **1.4** • Antonia Hirsch, *The Invisible Hand (after Adam Smith)*, 2009.

impaired adults ... The work 'makes visible' the situation of not seeing – and the implications of exclusion in a place of higher learning, with its promises of knowledge, social interaction and professional advancement."[26] On the other hand, *The Invisible Hand (after Adam Smith)* challenges the perception that the free-market economy (that is, one without controls) is unbiased, revealing how it selectively overlooks conditions that resist commodification. This work references the thinking of eighteenth-century Scottish economist Adam Smith, who, in *The Wealth of Nations* (1776), developed the idea of the "invisible hand" as a metaphor to explain how production and consumption will self-regulate based on the interests of buyers and sellers, an idea that serves as a cornerstone of capitalist markets. So, while the specific theme of these two works is distinct, in both works Hirsch brings the surveillance gaze as a system of control together with other systems of control, including the social and historical primacy of vision and of capitalism. Hirsch's works emphasize that surveillance structures of security – and the forms of looking they produce – do not operate in a vacuum, nor do they form a system unto itself; rather, surveillance structures intersect with, legitimize, and produce other forms of power and oppression.

Double Blind and *The Invisible Hand* destabilize assumptions about who has the right to see and who can claim full access to knowledge within regimes of surveillance visuality and its promises of security. In these works, Hirsch juxtaposes non-visual modes of perception (braille writing) with the technologies of seeing and surveillance (the domed mirrors) to challenge various social systems of control and regulation. These works resist common associations between seeing, access, and knowledge by denying most viewers the full sense of the message. As well, by asserting braille, with its association of touch and materiality, as central to the piece, she questions the presumption that the gaze, or the act of looking, is the primary or even dominant form of knowledge gathering. This is a creative strategy that curator Jan Allen defines in the exhibition catalogue for *Sorting Daemons* as "sightless watching."[27] Allen's idea of sightless watching is important in understanding the significance of Hirsch's projects to larger conceptions of the surveillance gaze because it moves access and knowing away from a purely visual act. Hirsch reveals the process of knowing via surveillance as not simply an act of seeing the visual representation of a surveilled scene; rather, she treats surveillance knowledge as a discourse, as a rhetoric of power and control that has resonances on and through the body and material forms.

Hirsch's two works reveal the constraints of surveillance structures by exposing the limits of ableist modes of seeing. These creative projects foreground the invisible privilege embedded in the presumptions of who gets to see through and interpret with the surveillance gaze. In so doing, they destablize the assumption that non-disabled or otherwise privileged bodies automatically possess full access to knowledge within systems of surveillance visuality, and, by extension, to the structures of colonial privilege that surveillance upholds. By restricting access to key information – particularly through the use of braille – *Double Blind* and *The Invisible Hand (after Adam Smith)* challenge the surveillance gaze's reliance on vision as a totalizing mode of control, exposing the incompleteness of visual access on which claims to security depend. Hirsch's creative projects therefore elucidate not only what the surveillance gaze purports to reveal, but also what remains inaccessible within its visual logic, fragmenting the promise of its comprehensive knowledge.[28] Using domed security mirrors that operate as a seeing device, the works ultimately recast the surveillance gaze – and its promise of security – as a contingent and coercive fiction, one that relies on the promise of total visibility while continually revealing its own points of illegibility.

Lewis's *Some will take more prodding. Others will be more difficult*, Nanibush's *Carrying*, and Hirsch's *Double Blind* and *The Invisible Hand* come together to question whether surveillance and its gaze really make people more secure. They ask: Who needs to be protected from whom? And under what interpretative contexts are safety, uncertainty, and acceptability assessed? Lewis's work illuminates how this power to interpret for the sake of security can produce insecurity for some; Nanibush's work shows us what happens to interpretation when the one doing the interpreting changes and thus the promise of security changes; and Hirsch's works address the invisible privilege and inequality inherent in the power to interpret from a surveillance gaze, as well as the limits to what the surveillance gaze can see. All four of these works highlight the subjectivities involved with seeing and, equally important, with interpreting what is seen, which, I contend, is a crucial method of accounting for who gets included or excluded within dominant narratives of Canadian-ness. These works point to the power and control that comes with being the one to provide meanings and explanations – both those produced in artworks and those produced by surveillance mechanisms. While these works invoke the surveillance gaze as a recognizable angle of viewing – whether that is through a representational vantage point, visual cues like blacked-out eyes, literal surveillance footage, or analog devices such as mirrors – they clarify that the surveillance gaze is not just about looking. It is also about a discursive organization of ideas, a rhetoric of access and monitoring implicitly or directly connected to control and to concepts of security that inform historical and ongoing narratives of settler colonialism, capitalism, white supremacy, and ableism in Canada.

The artworks in this section distort expected norms attached to surveillance formats and create slightly unfamiliar viewpoints, moving their viewers to rethink and perhaps challenge their own expectations of how and for whom systems of security operate within contemporary Canadian society. These works complicate the primacy of vision presumed to be at the heart of seeing surveillantly and highlight how the seeing of the surveillance gaze in all its complexity upholds systems of authority, inequality, and oppression. I argue that, by putting the viewer in the position of authority as the one who looks – or, in the case of Hirsch, the one who is thwarted from looking – these works point to the broader relationship between the surveillance apparatus and Canada's settler colonial project, a project that justifies its own authority under the pretext of security. Together, the works confront how feelings of security and comfort associated with surveillance and its gaze are unevenly distributed, and they ask whether surveillance truly increases security, whose security is assured, and under what interpretative parameters the assessments of safety, uncertainty, and belonging are made.

Seeing

Historically, surveillance has been framed as a form of seeing that flows from the vantage point of an omnipresent, ubiquitous "Eye of God."[29] This framing is informed by a belief that technologies of visual recording are inherently objective, disengaged, cold, and impartial. Such a belief denies the subjectivity of those who interpret surveillance capture, and it decouples surveillance logic from larger authoritative social and political systems. Despite the technological framing of surveillance as an empirical tool for truth-telling, the meaning of the images it produces is mediated by cultural, social, and institutional contexts. Viewers – whether security personnel, law enforcement, or the public – bring their own biases, assumptions, and expectations to the act of interpretation, often projecting suspicion, guilt, or innocence based on normative assessments of bodies, spaces, and activities in ways that reflect and reproduce broader systems of oppression. This subjectivity influences not only how surveillance footage is read but also how it is used to justify decisions, actions, or narratives affecting those being watched. The surveillance gaze constitutes a mode of seeing shaped by power structures that construct historical narratives.

To challenge perceptions that surveillance electronics remain detached from human interest and socio-political systems, the artworks in this section anthropomorphize surveillance equipment, giving it some human traits. Fragments of the human body become stand-ins for representations of surveillance technology and its gaze, highlighting the perceptions of who makes, interprets, and assesses surveillance images. By manipulating the idea of the supposed impartial "Eye" of surveillance – in some cases by animating a literal representation of a human eye – these artworks compel audiences to rethink how they understand surveillance technologies and the people or agents who use them. The works replace computers and electronics with the human form to make visible the imposition of surveillance technologies and the gazes they produce, which are often rendered invisible through their familiarity and ubiquity. Here, the artworks denormalize taken-for-granted

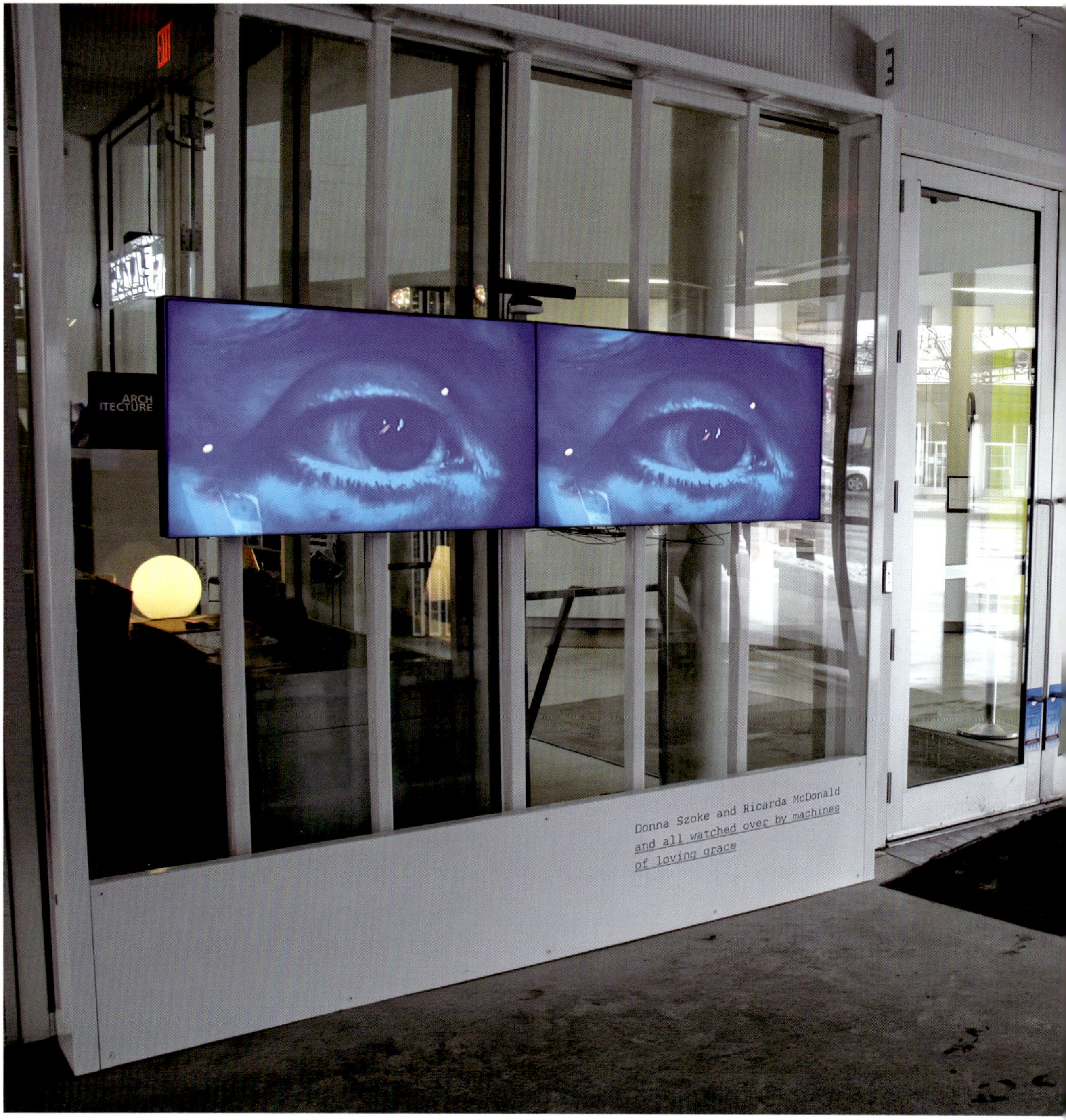

1.5 • Donna Szoke and Ricarda McDonald, *and all watched over by machines of loving grace*, 2012.

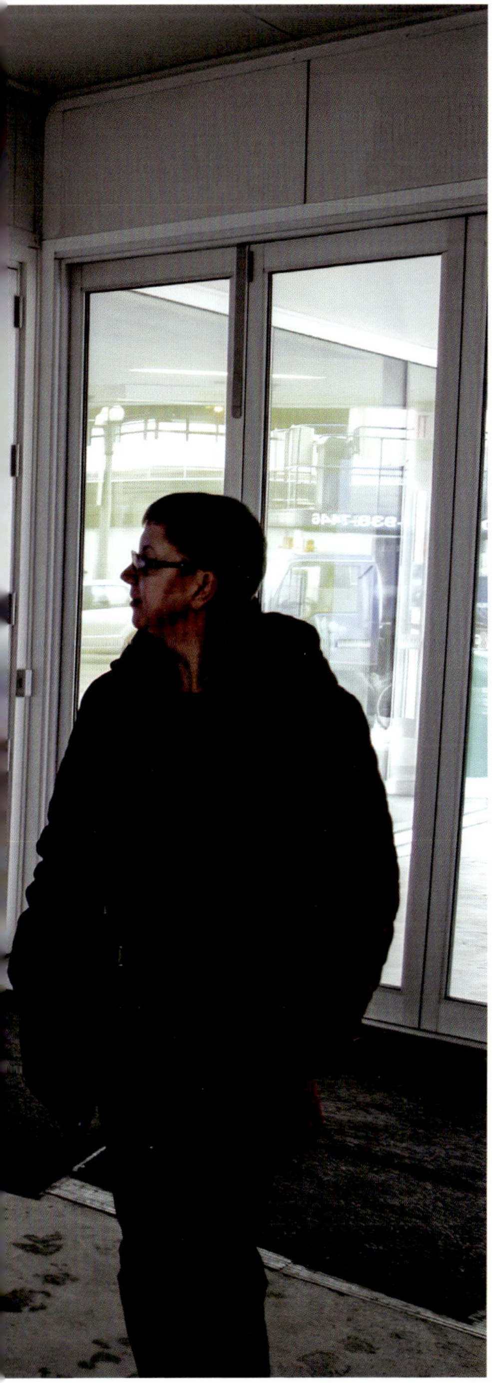

technologies that contribute to an "Orwellian condition of constant exposure and monitoring," whereby the surveillance gaze is embedded in contemporary life.[30] By playing with the bodily form of the technologies that produce the surveillance gaze, the artists in this section insist that the production of knowledge is always embodied, and therefore subjective, even when it comes to the seemingly mechanical, disinterested, disembodied surveillance gaze. In emphasizing the embodied subjectivity that underlies surveillance logic, they reframe the conventional relationships between those who are viewed, those who do the viewing, and the machines and technologies that permit such relations.

These projects foreground the technologies of surveillance in often satirical or over-the-top ways that magnify the processes of the surveillance gaze. A recurring theme in Canadian artworks that consider the technologies of surveillance is to create parallels between the eye of the surveillance camera and the eye of the human. One of these projects, Donna Szoke and Ricarda McDonald's *and all watched over by machines of loving grace* (2012, fig. 1.5), plays with expectations around bodies, looking, and technology. Two large monitors, each displaying a close-up view of an unblinking eye that stares out at the audience, sit side-by-side. Yet, these two eyes – one in each monitor – do not form a pair; rather, both screens represent a left eye, disrupting expectations of a matching set. As people walk by the screens, the two eyes fix on them and follow their movements, just as a motion-detection surveillance camera would. Drawing a connection between the eye of the camera and the eye of the human, this piece undermines the presumed objectivity of the technologies of surveillance by exposing the subjectivity and limits of those who (often invisibly) control them.

To further draw out parallels between the eyes of the camera and the eyes of the human, a video in an adjacent room plays a continuous loop of an eye that blinks in slow motion, highlighting the moments when vision is obscured by the eyelid, again underlining the limits of watching. According to Szoke and McDonald, the installation "emphasizes the human element of surveillance – the blink – that implies an inherent failure if compared to machine surveillance."[31] For the artists, then, the work juxtaposes the all-seeing, unrelenting gaze of the machines of surveillance with the inadequacies of human viewing for surveillance purposes, because the blink represents an absence in that relentless stream of information gathering. I think, too, this juxtaposition reveals the inextricable interconnections between these two forms – human and machine – and interrogates the limits of each, independently and together. It raises questions of who gets to occupy the invisible space behind the camera and what ambitions they project onto the targets in front of the camera. What interpretative contexts and authoritative systems guide the way they understand the images of the surveillance gaze?

The title of the piece is taken from a 1967 Richard Brautigan poem that describes a vision of a loving and caring gaze of humanity. The artists use this reference to a utopian version of the gaze as an overt critique of what they see as the uncaring and ruthless surveillance gaze, which has been turned against humanity through its obsessive concern with property crime and consumerism.[32] In this conception of the punitive surveillance gaze, prioritizing property, interpreting crime, and upholding capitalism, we know that certain bodies are revealed and interpellated as inherently suspicious and in need of constant monitoring. Others, aligned with dominant social norms, retain their privileged position as watchers, tasked with protecting spaces of home and business from perceived encroachments. This dynamic reinforces structural systems of racism, classism, cisheteropatriachy, ableism, and settler colonial notions of belonging. While the artists' reference to a utopian version of the gaze sets surveillance and humanity in opposition, I argue that the work simultaneously inserts humanness into the surveillance gaze by foregrounding the subjectivity, and thus discernment and discrimination, behind it.

The personification of the surveillance gaze through the literal gaze of a disembodied eye is not unique to Szoke and McDonald's project.[33] Artist duo Jennifer Marman and Daniel Borins also use this strategy in their project *Google* (2009, fig. 1.6), a sculptural installation composed of two large "googly eyes" originally mounted on an interior wall at the Martin Prosperity Institute in Toronto. Partially hidden

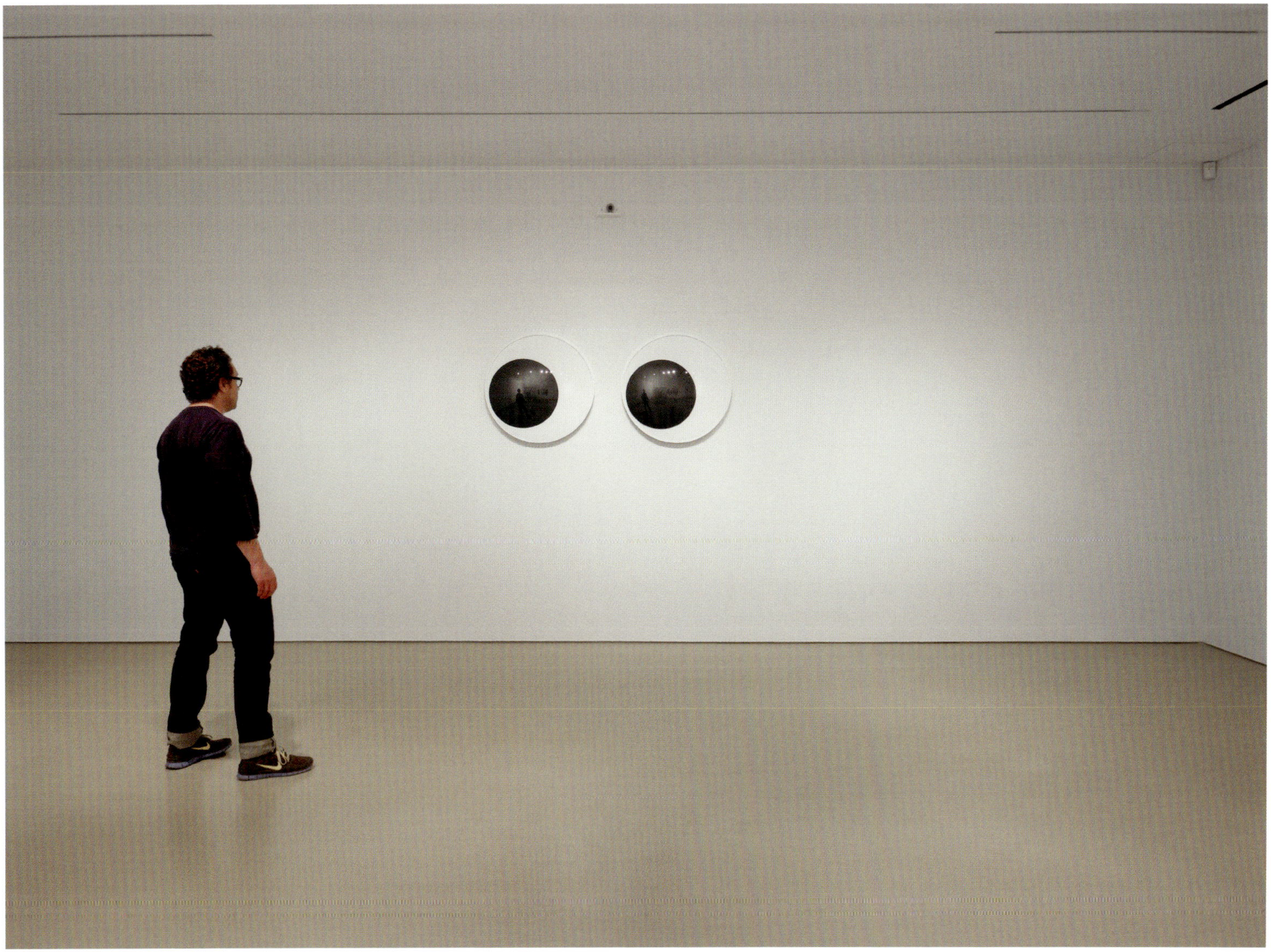

1.6 • Jennifer Marman and Daniel Borins, *Google*, 2009.

cameras inside the sculpture watch, track, and record people as they pass by. The large black iris – with the lens of the camera noticeable behind it – follows movement, adding a potentially menacing element to the humour offered by the googly eye form. As a double entendre, the title of the piece acknowledges the ways in which internet search engines – the most prominent of which is Google – collect and collate user data for ambiguous reasons.

In this piece, however, the placement of the camera inside a recognizable yet unexpected form – the googly eye – defamiliarizes the role of surveillance in contemporary society as supposedly predictable, mundane, and commonplace. It makes the imposition of the surveillance gaze awkward and hypervisible, the camera rendered large and cartoonish. As art historian Gabrielle Moser writes of the piece, "these are not the eerily invisible surveillance devices of George Orwell's *Nineteen Eighty-Four*. Instead, Marman and Borins offer systems for seeing that are clunky and awkward, prone to glitches and meltdowns caused by their internal logics."[34] Using humour as a device, the work simultaneously points to the prevalence of surveillance systems and the limits of their mechanisms, showing that surveillance technology will inevitably be prone to glitches and meltdowns no matter how elegant the technology appears. As well, the satirical bent of this work conflates camera surveillance with data surveillance to make both forms visible and discernible to audiences in order to expose the embeddedness of data extraction by governments and tech corporations such as Google in both our physical and digital everyday lives. In this way, *Google* critiques the normalization of and connections between the surveillance gaze and colonial capitalism, and raises urgent questions about the trust we place in entities that benefit from the commodification and control of our personal information.

The glaring presence of the googly eye as a surveillance device illuminates the ubiquity of surveillance in the contemporary environment and highlights the mass political acceptance of this form of monitoring as essential for public and national security. The surveillance camera is a recognizable and visible form of the surveillance gaze that has been integrated into Canadian society to the extent that it is now widely accepted as an inevitable part of daily life, particularly since 2001, when certain populations who were historically excluded from the monitoring and criminalizing lens of surveillance became more aware of being caught in its gaze. Yet, this work contests the idea that the camera's presence and gaze are truly innocuous. It asks whether state and corporate encroachments on individual privacy serve their proclaimed purposes, or whether they reinforce existing authoritative systems under the guise of security and efficiency. The emphasis on the eye of the camera underscores the presence of a person – or a subjective system – whose assessments and decisions about surveillance footage are shaped by entrenched systems of oppression; it reveals how surveillance is never neutral but instead operates through power-laden judgments about whose behaviour is deemed deviant or threatening. The comical form of the cartoon eye that tracks people as they walk past asks the audience whether they care if and how much their privacy is breached, testing the trope "I have nothing to hide." In doing so, *Google* reveals what it means to be caught indiscriminately in the net of surveillance; it produces the discomfort of being targeted by the policing apparatus as a "person of interest" within the dominant gaze of the Canadian state and international corporations. With this work, Marman and Borins reappropriate camera surveillance tactics to question their very ethics and intent.

Both *and all watched over by machines of loving grace* and *Google* anthropomorphize the gaze of surveillance machines by repackaging the camera lens as a human eye. So too, they prompt audience members to ask, if I am the one being watched through this lens of surveillance, then who or what is behind the camera making decisions and assessments? And what interpretative contexts or assumptions inform those assessments? Often, conventional structures of surveillance render the watcher invisible, hidden behind the supposed objective lens of the surveillance machine. Those who are watched, on the other hand, are made markedly visible, carefully observed and analyzed, each movement, action, body, and characteristic carefully tracked and evaluated for uncertainties, suspicions, or threats. Of course, this monitoring differentially affects specific bodies, actions, and locales, as the interpretation of surveillance data is deeply embedded within and shaped by intersecting regimes of oppression within the settler colonial context of Canada that dictate whose behaviours are perceived as suspicious, deviant, or threatening and whose are deemed normative and deserving of protection. These oppressive frameworks not only influence how surveillance is enacted, but also determine the social and political consequences for those subjected to its gaze. Works like those of Szoke and McDonald, as well as Marman and Borins, cast into view the eye of the beholder that sits – invisible and hidden – behind the surveillance apparatus and assesses those caught in the glare of its gaze. They make visible that which is often invisible: they make the production and interpretation of a gaze by the surveillance camera apparent and present.

Chicago-based artist Pat Badani does not use just the seeing eye in her work: she uses the entire face to make a similar critique of the purportedly objective, neutral gaze of surveillance technologies, in her case by emphasizing the way in which the surveillance gaze's anticipated evaluations generate an embodied sense of being viewed. Badani produced *make-A-move* (2015, fig. 1.7) while she was in residency at Agence TOPO, an artist-run centre in Montreal that provides the space and time for artists to develop artworks and writings on digital culture.[35] In 2015, *make-A-move* was included in the Art Souterrain Festival, also in Montreal. This project features two close-up videos – one of a white-presenting man's face and one of a white-presenting woman's face – installed on the side of two large red boxes in the middle of a busy underground pedestrian thoroughfare. As people walk in front of the digital images, the eyes of the figures follow them. The faces do not react to the movements of passers-by only with their eyes; they also react with their facial expressions. In response to people and their actions, the features of the faces alter into expressions of suspicion, fear, surprise, or reproach. Effectively, they pass judgment on the passers-by as an explicit way to evoke in them a heightened sense of self-awareness. The surveillance faces racialized as white highlight the racial and colonial inequities of the gaze within Canada, equating those who watch and appraise with dominant forms of whiteness in the settler colonial context.

The press release for the Art Souterrain Festival, curated by Carmit Blumensohn with Raymond Cantin, describes how Badani "playfully engages questions about

1.7 • Pat Badani, *make-A-move*, 2015.

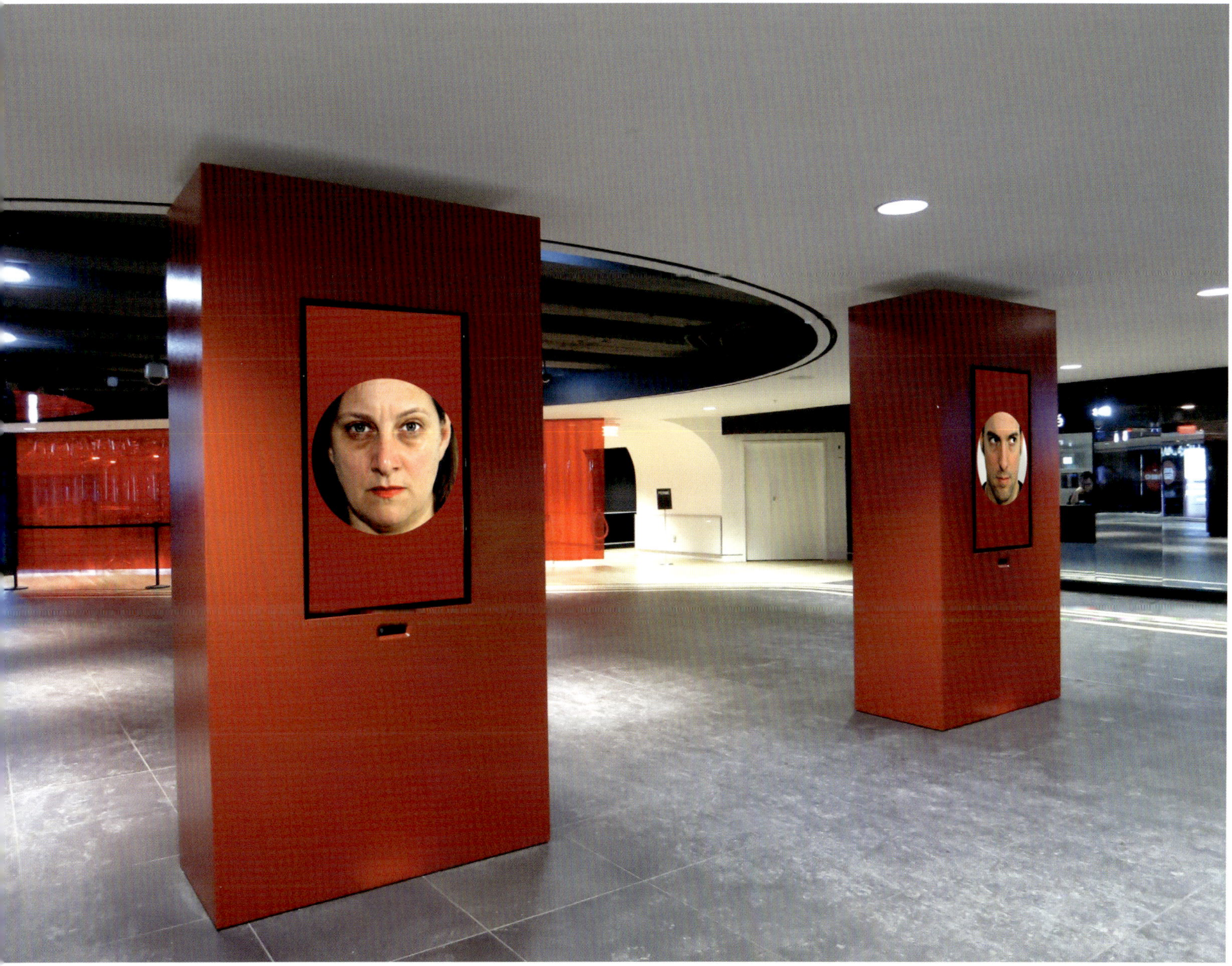

mutual security and also points to the psychological, personal, and collective cost of electronic surveillance mechanisms that track our data and traffic."[36] *make-A-move* explores how the awareness of being under the watchful eye of surveillance that scrutinizes every move can promote self-regulation and potentially feelings of insecurity or uncertainty about the way we go about our everyday lives, particularly in public spaces. More than just examining the relationships between surveiller, surveilled, and surveillance technologies, this work exposes the unseen psychological, mental, and emotional strands that underlie and thread through these relationships. This installation lays bare the acts of looking or being looked at and also the rubric against which what is observed gets processed, categorized, and evaluated – here, explicitly, the invisible norm of whiteness. It highlights the dominant and authoritative gaze within settler colonialism from which categorizations of belonging and inclusion are determined. This information-gathering and assessment process is often hidden from view, but Badani puts it on display with *make-A-move*, which equates information analysis derived from the surveillance gaze with social judgments and interpretative contexts that elevate whiteness as the arbiter of and benchmark for inclusion.

In these three projects – *and all watched over by machines of loving grace*, *Google*, and *make-A-move* – the artists make a connection between the seeing of surveillance technologies and the human mechanics that organize them. Szoke and McDonald, Marman and Borins, and Badani all play on historical associations of the camera as an extension of the human eye by making this metaphor literal. Specifically, each work uses literal depictions of the human body as embodied forms of the surveillance gaze so that viewers may be alerted to and reminded of the people and organizations that manage the supposedly neutral and unbiased "Eye" of surveillance watching. These works both centre the seeing of the surveillance gaze and demonstrate that this gaze is not an isolated visual phenomenon: it is inextricably connected to larger social and political systems.

Surveillance systems are often framed as dispassionate mechanisms of security and efficiency, yet they are in fact deeply shaped by the biases, assumptions, and cultural logics of those who design and operate them and who claim authority within the Canadian project of settler colonialism. This recognition reveals that surveillance does not simply record reality but actively constructs it, selectively highlighting certain behaviours, bodies, and spaces while obscuring others. Acknowledging this constructedness opens space for critical interrogation: Whose vision is encoded in surveillance technologies? Who is made visible and under what terms? By putting the familiar (the surveillance camera) in an unfamiliar form (the humanized electronic eye), these three artworks highlight the intensity, prevalence, and normalization of the surveillance gaze, and encourage audiences to grasp the connections between surveillance technologies and the dominant and authoritative interpretative contexts that govern and structure them.

Being Seen

The artworks in the previous section reveal the subjective structures that govern who is seeing through the surveillance gaze and the interpretative contexts that frame how that seeing is assessed. The creative projects in the present section focus on the other side of the lens: the one caught in the surveillance gaze. The feeling of being seen by the surveillance gaze intersects profoundly with the ongoing realities of settler colonialism in Canada, where surveillance systems and their gaze operate not merely as neutral technological apparatuses but rather as tools of colonial governance and social control. Surveillance systems – particularly those rooted in state and corporate structures – often abstract individuals into data points, rendering them legible through dehumanizing tactics shaped by racism, colonialism, cisheteropatriarchy, ableism, and other forms of systemic inequity. When it disproportionately targets marginalized populations, marking racialized, gender-nonconforming, disabled, and economically disadvantaged bodies as threats or anomalies, the surveillance gaze selectively amplifies visibility. This section examines artworks that challenge differential visibilities that produce certain subjects as inherently suspicious, risky, or deviant while rendering others as normative and invisible, as the works question authoritative looking from the position of the one who is watched.

In that they interrogate the uneven levels of visibility applied to the watched, the artworks in this section address what communications and surveillance studies scholar Torin Monahan identifies as a key question asked by critical surveillance art: "What kind of subject is the subject of surveillance?"[37] Embodiment plays a key role as an artistic strategy here, as the artworks bring the art visitor into the work itself in creative and distinct ways to encourage visitors to feel the moment they become implicated in the surveillance structure. These artworks highlight that the gaze is not purely an intellectual, disembodied, or autonomously visual experience; rather, as sociologist Gavin Smith articulates, "bodies have a key role to play in practices of surveillance ... both in terms of how they watch and how they are watched."[38] The two artworks discussed here bring audiences into the piece as targets of surveillance, providing opportunities for them to occupy the position of the watched – the surveilled – and to examine their own understandings of what this space means to them. In so doing, these art projects force the audience to rethink and "refeel" the experience of being seen through and constructed by the surveillance gaze, a gaze that carries out on them the vision of control offered by the Canadian settler colonial project.

The strategy to implicate the art audience into the surveillance aspects of the piece itself is one that Ursula Johnson employs in her installation *Mi'kwite'tmn (Do You Remember)* (2014).[39] The first two of the three distinct spaces of the installation involve unconventional representations of traditional Mi'kmaw basket weaving, either as sand-blasted diagrams on vitrines or as deconstructed basket-like objects. In the third space, Johnson herself makes a performative gesture of preparing wood for baskets in real time. Her performance in this part of the exhibition begins the process of preparing a log using traditional knowledges passed down to her from elders in her community. As the show's curator, Robin Metcalfe, explains, this work addresses "the impact of colonialism on aboriginal material and linguistic culture, and challenges the museological and ethnographic frame placed upon it."[40]

Importantly, the performative aspect of this artwork relates to Johnson's longer engagement with the concept of *netukulimk* – a Mi'kmaw term that loosely refers to the idea of "self-sustainability" – and with how it might apply to basket weaving.[41] In this work, Johnson emphasizes the "mutually beneficial and respectful relationships" inherent in the traditional processes of making and in learning the necessary skills, which, in her words, reject the "disconnect between people from my generation in regards to our relationship to the natural world."[42] So too, as curator and Indigenous and Canadian studies scholar Rachelle Dickenson writes in the exhibition catalogue, this performance intends to create "a space of *rapprochement*: building bridges between knowledges, developing relationships that confront stereotypes, reaffirm community connections and perhaps create new ones."[43]

Mi'kwite'tmn (Do You Remember) juxtaposes Johnson's nurturing of respectful relational connections and intergenerational knowledge through basket weaving with a series of imposing surveillance cameras mounted in the gallery space that track and monitor the visitors who watch Johnson work. As she shaves, pounds, and splits a log in order to produce the materials necessary to start the basket-weaving process, the surveillance cameras record visitors upon entry.[44] The addition of surveillance monitoring in this space, according to Johnson, is meant to provoke a feeling of being watched. She explains:

> If non-indigenous viewers come into the space, I need to create a way of feeling that they're being watched. Because it's very different when your cultural context comes into play ... For many of us who exist within marginalized groups in society, if someone is standing in the corner of the room, then you're being watched. But somebody else might be thinking, "Oh, this person [in the corner] is here to help me! They're not here to keep an eye on me and make sure I don't do anything wrong." So I wanted to create that feeling of being watched and not being trusted by using digital media by having these surveillance cameras in the space.[45]

For Johnson, the surveillance aspect of the work implicates the audience member, particularly the non-Indigenous audience member, as the one not to be trusted. By juxtaposing the act of preparing for basket weaving – anchored in respectful and mutually beneficial relationships – with the presence of surveillance cameras, Johnson defines surveillance, and thus the anti-Indigenous, settler colonial apparatus that often wields it, as antithetical to community building.

Mi'kwite'tmn (Do You Remember) uses surveillance strategies to implicate people who may conventionally remain outside of direct systems of surveillant policing in the settler colonial context of Canada. This work directly challenges histories of white supremacy and colonial settlement in Canada in its relation to surveillance logic and its gaze. By subjecting non-Indigenous visitors to the experience of being surveilled, and to the feelings of incrimination attached to that experience, Johnson positions them as outsiders within an Indigenous-controlled space. By flipping the colonial script, Johnson's work raises several questions: Who is conditioned to feel welcomed or unwelcomed in museological and ethnographic spaces that function as microcosms of national spaces? What hidden norms are protected by conventional practices of surveillance? Who is most often targeted by the surveillance gaze? With *Mi'kwite'tmn (Do You Remember)*, Johnson highlights the racial, colonial, and systemic inequities that inform whose bodies get caught in the nets of surveillant policing and whose generally do not. The artwork refocuses the lens of historical and ongoing colonial systems that have violently surveilled and policed Indigenous people and communities as part of the foundational logic of white settler Canada.

Another artwork that targets the art audience with the surveillance gaze as a mode of critique is David Rokeby's *Gathering* (2004, fig. 1.8). Rokeby has produced a number of works that visualize the act of viewing and compel a certain level of participation from the audience in becoming the subject of the surveillance gaze. His artwork *Gathering* is a large-scale video installation that records moving images of people outside the gallery and then projects recontextualized versions of these images inside the gallery. Commissioned by the Art Gallery of Hamilton in southwestern Ontario for the São Paulo Biennial in 2004, this artwork positions a

1.8 • David Rokeby, *Gathering*, 2004.

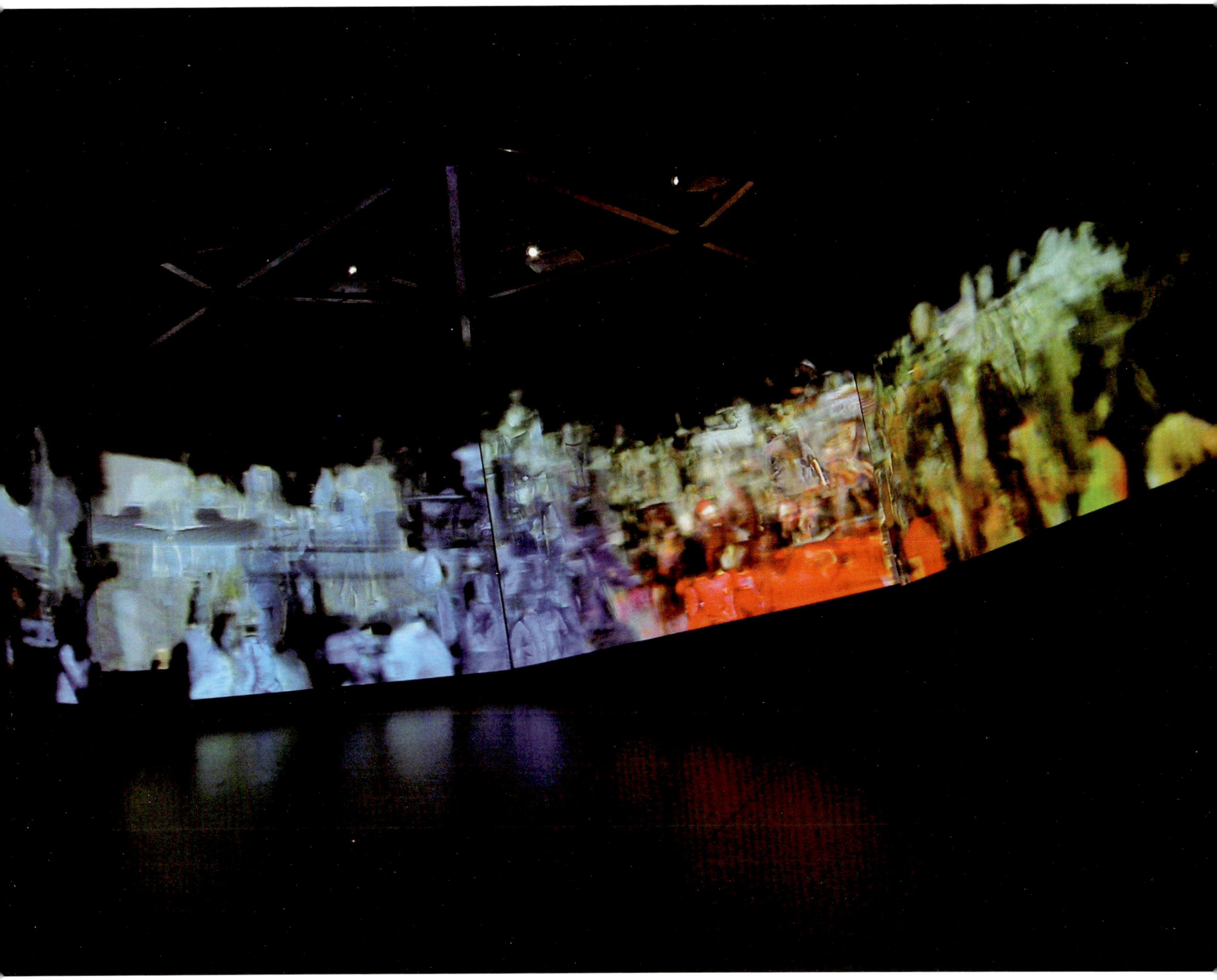

camera to overlook the exterior of the gallery, from which vantage it captures moving images of people on the street. Then, a software algorithm separates the people in the images from their surroundings, and analyzes and groups them by colour. Rokeby explains the process of analysis and grouping like this: "Flesh toned areas are separated from clothing and the clothing is divided into areas of like colour. These moving patches of colour (heads, t-shirts, jeans, coats, hands) are then sorted across the screens."[46] These segmented portions of the captured images of people are projected onto a large screen inside the exhibition space, organized according to either vertical or horizontal patterns of matching colour tones. The resulting effect is a moving collection of abstracted, prismatic shapes – still noticeably derived from images of people – that periodically reorganize based on cycling criteria.

To be sure, setting skin tone and colour as the metric by which to separate and organize bodies points to historical and contemporary processes of racialization and racial profiling central to surveillance logics and the settler colonial project in Canada. These concerns are brought to bear in studies that examine how facial processing technologies (FPT) are able to identify and track people based on their skin tones. For instance, the Gender Shades project out of the MIT media lab examined three established corporations (IBM, Microsoft, and Face++) that offer artificial intelligence to assess the gender and race of subjects. Their research revealed the deeply gendered and racist biases of these technologies, with much higher reliability when the subjects were male and had lighter skin tones.[47] As the researchers of the Gender Shades project articulate, "automated systems are not inherently neutral. They reflect the priorities, preferences, and prejudices – the coded gaze – of those who have the power to mold artificial intelligence."[48] Yet despite their identified inaccuracies, these FPT algorithms are being deployed by law-enforcement units and in air travel security, reproducing surveillance patterns that reflect existing societal prejudices. As a group of international researchers studying the social implications of artificial intelligence warn, "the reality of FPT deployments … reveals that many of these uses are quite vulnerable to abuse, especially when used for surveillance and coupled with predatory data collection practices that, intentionally or unintentionally, discriminate against marginalized groups."[49] The interpretative context of Rokeby's *Gathering*, then, is inseparable from the historical and ongoing development and application of surveillance systems that track and mark certain bodies – however inaccurately – and replicate societal biases that reproduce colonial, racist, and cis-heteropatriarchal perceptions of the body.

This work by Rokeby uses recorded images of people who are not necessarily willing participants. While it may feel as if this non-complicity of unsuspecting people is nefarious in some ways, *Gathering* merely replicates what security and surveillance cameras across Canada (and elsewhere) do every day. Rokeby simply holds a mirror to a facet of Canadian life that takes place, regardless of whether people know about it or not, in public and private spaces around the country. With this work, Rokeby does not focus on the body in its entirety; rather, he dismantles the bodily form and reconstructs its pieces in different ways based on randomized algorithms. This process dehumanizes or unhumanizes the different elements, because the artwork is focused on people not as whole entities but as aggregates to be mined and categorized to generate particular forms of data and understanding. This approach reflects Hier's observation that "the proliferation of information and data gathering techniques ... operate[s] to break the human body into a number of discrete signifying data flows."[50] Rokeby's process of data mining bodies based on colour highlights the racism key to surveillance and its algorithms. African American studies scholar Ruha Benjamin asserts that "invisibility, with regard to Whiteness, offers immunity,"[51] while communities of colour are monitored and surveilled. In his work, Rokeby points to the ease with which those who control surveillance technologies and systems capture and recontextualize images according to their desires. He shows how the algorithms used to categorize people reflect the dominant interests and assumptions of the people and societies that structure the surveillance and algorithmic systems.

The creation and direction of algorithms, as internet and digital justice scholar Sofia Noble argues, "reinforce oppressive social relationships and enact new modes of racial profiling" because the formulations that generate "automated decisions are made by human beings."[52] *Gathering* emphasizes that the surveillance gaze is not neutral or objective; the work demonstrates how simple it is for interested parties to direct algorithms to distinguish any particular feature as the central defining element in the categorization and understanding of an image, person, or activity. This recontextualization is key because it highlights how the surveillance gaze can be reimagined in a variety of scenarios, altering the way people understand or interpret images. *Gathering* prompts audience members to ask questions such as, in what subjective, meaning-making ways am I visible or invisible? Visitors viewing this artwork become the central visual components of it; thus, the act of seeing is also one of being seen. The same person occupies both positions – watcher and watched – and the level of surprise or discomfort in witnessing one's own visibility can be directly correlated to the level of presumed surveillance that one experiences in everyday life. Gallery visitors confront their own level of comfort as one who sees versus one who is seen, along with the social, political, and cultural implications that come with each position.

I organized this section around the theme of being seen because this aspect is important to the way these artworks function. Yet it is not where their emphasis ends. The artworks take the action of being seen and make it simultaneous with the act of seeing. Whereas in the previous section, "Seeing," the artworks create an encounter between art visitors and the seeing "Eye" of surveillance looking, in "Being Seen," the works implicate viewers as objects and subjects of the surveillance gaze, prompting viewers to contemplate the vulnerabilities, power, and control inherent in each of these roles. Johnson and Rokeby developed projects that force audience members to participate – willingly or unwillingly – as subject matter and visual components of the very art they are there to take in. These artists reimagine the surveillance gaze – often inextricably linked to the visual and visible – by actively pushing engagement into the realm of embodiment. Again, the way they encourage this bodily experience is by making the viewer who sees simultaneously the one being seen: the subject and object of the artwork. This strategy reframes expectations of both the surveillance gaze and the artwork itself by questioning the presumed separation of each position, frustrating so-called critical distance in order to uncover the dominant desires that underlie surveillance motives. Both works actively position art visitors not as viewers of a distanced artwork with subject matter discrete from their own experiences or bodies, but as viewed objects central to their own viewing.

These creative projects challenge the control of the surveillance gaze by shifting attention from the technological apparatus and its power to those who experience, organize, and interpret its effects. By critically reflecting on what it means to be seen – who is seen, how, and under what conditions – this strategy exposes the unequal distribution of visibility and the political stakes of representation. It reveals that visibility under surveillance is not a matter of mere observation, but also of social control, shaping how certain bodies are read, judged, and acted upon, reducing them to data or risk categories within systems structured by racism, colonialism, and other forms of oppression. In foregrounding the emotional, embodied, and political dimensions of being seen, the artworks unsettle the seeming neutrality of the surveillance gaze, and interrogate the systems and logic that underscore it. Both highlight the dominant paradigm that targets certain populations with the surveillance gaze, and the role that data and information plays in further bolstering and affirming systems of power, control, and authority.

The artworks in this chapter centre the surveillance gaze as an established and recognizable form of looking, then distort it as a strategy to encourage art visitors to question the systems that normalize this form of looking in the first place. Addressing the broader conditions of surveillance via its identifiable aesthetic characteristics, these creative projects engage not just with surveillance as a form of seeing but also with surveillance as a social practice – as a form of understanding inextricably linked to structures of power that shape history. They mimic and alter forms of looking attached to the surveillance gaze to reveal its role in reproducing the social hierarchies by which the disciplinary power of the Canadian settler colonial state is organized and maintained. Lewis's *Some will take more prodding. Others will be more difficult*, Szoke and McDonald's *and all watched over by machines of* *loving grace*, and Marman and Borins's *Google* illuminate capitalist extractions and the bodies that are implicated in these processes; Nanibush's *Carrying*, Johnson's *Mi'kwite'tmn (Do You Remember)*, and Rokeby's *Gathering* highlight the legacies of colonialism and ongoing white supremacies; and Hirsch's *Double Blind* and *The Invisible Hand* question ableist expectations related to meaning making through surveillance access. These artworks take up the aesthetic characteristics of the surveillance gaze to critique its discursive and material implications.

These works demonstrate that the surveillance gaze is not neutral. They do so by blurring the distinctions of watcher and watched and by putting the bodies of the surveilled in direct, immediate, and visible relation to the bodies (or body parts) of the surveillers. In "Security," I explored how the creative projects problematize who has access to surveillance looking, who gets to produce meaning from what they see, and who gets to feel safe and secure. In "Seeing," I discussed how they centre the human body to challenge perceptions of surveillance looking and its gaze as neutral or disembodied. Finally, in "Being Seen," I showed how the artworks implicate the art audience in order to merge positions of seeing and being seen and to challenge the unseen authority of the surveillance gaze. To be sure, there is a marked unevenness between the agents of surveillance and those who are watched, and each of these works problematizes these expectations in the context of belonging and the Canadian state.

By denormalizing the embeddedness of the surveillance gaze in everyday life, these projects build on and contribute to contemporary reinterpretations of Foucauldian disciplinary power that highlight how the surveillance gaze functions not merely as a tool of observation but also as a fundamental mechanism in the architecture of settler colonialism. As criminologist Jeffrey Monaghan articulates in relation to nineteenth-century colonial policies against Indigenous people and nations in the Canadian Northwest, "colonial authorities develop systems of identification, categorization, and enumeration that become surveillant infrastructures for the biopolitical management of colonial populations."[53] By rendering national populations as subjects of suspicion and control, surveillance practices continue to facilitate the dispossession of peoples from the land, the suppression of resistance, and the oppression of positionalities assessed as threatening to the norms of the state. The deployment of this surveillance logic in the form of the gaze – a surveillance visuality that brings forward angles of watching and being watched in the service of policing, monitoring, and knowing – is the central method by which the seeing of surveillance becomes attached to the systems of power and control of settler colonialism. Individually and as a group, the nine creative projects explored in this chapter expose the surveillance gaze in Canada as a disciplinary mechanism of the settler colonial state, demonstrating how its visual regimes are fundamentally entangled with white supremacy, racism, capitalism, and ableism. By centring the surveillance gaze as a critical and creative point, these projects underscore the necessity of understanding the surveillance gaze and its logic not as an isolated phenomenon with intrinsic coherence but, importantly, as deeply intertwined with the ongoing processes of colonization and settler state formation.

2 Policing

Surveillance is a technology of normalization that identifies and discourages the cultural expression and behavior of social formations that differ from those of the dominant, and thus chills any public display of difference.

John Fiske, "White Watch"

Surveillance systems and policing strategies are deeply intertwined with the ongoing structures of settler colonialism, functioning as mechanisms to assert and maintain state control. The artworks in this chapter situate surveillance structures in relation to policing activities, and I examine this relation to consider the implications of settler colonial histories in Canada. Each of the works incorporates different characteristics of policing as an authoritative strategy that helps to define and classify who is brought into or excluded from national inclusion. The artworks interrogate the ways in which policing strategies operationalize surveillance logic to protect the interest and goals of the project of Canada, and they offer new ways to think about the larger and often more complicated structures of control, privacy, in/visibilities, and access that align with ongoing forms of settler colonialism. Together, the eight artworks in this chapter illuminate the role of surveillant policing in Canada to seek out, criminalize, suppress, and eliminate practices, people, and communities that challenge the dominance of settler colonialism, with its intersecting systems of white supremacy, racism, capitalism, cisheteropatriarchy, and ableism.

While collusion between state and non-state surveillance actors characterizes contemporary surveillance systems, surveillance as a strategy of policing and control in creating and maintaining nation-states has a long history.[1] Within that history, however, 2001 and the War on Terror amplified and intensified established surveillance systems, immensely extending their reach. Certainly, the general public – especially those who were previously invisible to surveillance structures because of privileges based on race, gender, class, religion, ability, and/or sexual orientation – became more aware of surveillance systems as part of the "new normal" of post-9/11 life in Canada.[2] As surveillance studies scholars Kevin Haggerty and Richard Ericson assert, "hierarchies of visibility are being levelled, as people from all social backgrounds are now under surveillance. While surveillance has not eliminated social inequalities, certain groups no longer stand outside the practice of routine monitoring."[3] It is true that the surveillance net cast in the post-2001 era is larger, but it is essential to understand that this security environment continues to reinforce historical inequities that are contemporary and ongoing. As communication and surveillance studies scholar Torin Monahan explains, "it must be recognized that a host of surveillance functions are reserved for those who threaten the status quo, principally those classified as poor or marked as Other."[4]

Certainly, the expanded forms of surveillance that the War on Terror brought about are not distinct from – nor do they work in opposition to – surveillant policing logic present within settler colonialism. The imperial project of the War on Terror is inextricably linked to the settler colonial project of Canada (as well as to that of other nations involved in the Coalition of the Willing).[5] It is an extension of the colonial and imperial logic that expands the nation-state outside its physical borders to territories the state deems dangerous. Inside state borders, the War on Terror has expanded government and corporate rights to claim access and deny privacy to the state's citizens/inhabitants, and to further codify who is included in or excluded from settler colonial claims to the land. Critical Muslim studies scholar Shaista Patel says of the post-9/11 security campaign in Canada that government policies – which include Bill C-36: The Anti-terrorism Act, and Bill C-42: The Public Safety Act – function as "juridical discourses" that "provide certain legitimacy to the racist rationalities underpinning the project of making the Canadian nation a white space."[6] Patel discusses how new laws and new practices of surveillance intensified the criminalization and scrutiny of Muslims of colour in the post-9/11 Canadian context specifically, yet the logic behind surveillance practices more broadly both draws from and furthers histories of white supremacy and resource extraction that define Canada as an ongoing settler colonial project. As media scholar John Fiske succinctly articulates, "surveillance is a technology of whiteness."[7]

In Canada and beyond, there continue to be vast disparities in how certain individuals and groups are targeted and policed through surveillance. In this chapter, I use the term "surveillant policing" to highlight the logic of surveillance that comes to bear on policing. The term draws on the idea of "high policing," originated by criminologist Jean-Paul Brodeur and defined by public policy scholar Fahad Ahmad and criminologist Jeffrey Monaghan as "surveillance practices that target perceived or real forms of political opposition to the ideological hegemony of the state and its capacities for self-preservation and accumulation of power."[8] As I use it, "surveillant policing" also draws on activist Mariame Kaba's discussion of abolitionist organizing, where she defines policing as "an entire system of harassment, violence, and surveillance that keeps oppressive gender and racial hierarchies in place."[9] With these considerations in mind, I use surveillant policing to highlight the interconnections between surveillance logic and policing activities, which are mechanisms of control mutually motivated to manage racialized, gendered, ableist, cisheteropatriarchal, and classed hierarchies in the construction and maintenance of settler colonialism under the guise of public safety and security. Surveillant policing in this context serves to reproduce systems of power and domination, and to eliminate those who challenge the social and political order that upholds the authority of the Canadian state project.

To be clear, surveillant policing does not refer strictly to the institutional mechanics of monitoring and controlling by police or military forces. My use of the term gestures toward those larger and often more complicated structures of control and authority that align with historical forms of settler colonialism. Although the agents of surveillance structures are at times recognizable and easily identifiable, surveillant policing also operates in diffuse and systemic ways that extend beyond a single institution or actor. State violence is not limited to agencies of law enforcement; rather, as historian of Black feminisms Robyn Maynard explains,

> state violence can be administered by other institutions outside the criminal justice system, including institutions regarded by most as administrative, such as immigration and child welfare departments, social services, schools and medical institutions. These institutions nonetheless expose marginalized persons to social control, surveillance and punishment ... These bureaucratic agencies, too, have the repressive powers generally presumed to belong only to law enforcement. They can police – that is, surveil, confine, control and punish – the behaviour of state subjects.[10]

These multiple forms of policing intersect with social, cultural, political, and historical elements, and are informed by presumptions about people and communities, and by expectations about who is permitted to access particular spaces and locales. The underlying rationale of surveillant policing is informed by presuppositions about criminality, terror, and geopolitical systems, and seeks to present, consume, and regulate certain bodies, spaces, and activities in the name of the settler colonial order.

In this way, the policing opportunities offered by structures of surveillance since 2001 – which, again, are not always tied to obvious policing agencies – parallel, build on, and extend pre-existing modes of coloniality, control, and oppression. In other words, surveillant policing after the events of September 11th offer new forms of colonialisms – ones that activate the logic and rhetoric of terror and security but deny the uneven and unequal ways in which these forms are applied and operate. For instance, Maynard's research on policing and anti-Blackness in Canada confirms that "the over-surveillance, not Black proclivity to crime, is the leading factor in the disproportionate arrest of Black youth."[11] Monaghan's study of the development of interagency surveillance networks for the purpose of monitoring and surveilling "violent Islamic extremism" attests to the increase in racial profiling of Muslims of colour after 2001.[12] Monaghan and sociologist Andrew Crosby's account of the RCMP investigation Project SITKA as an extension of the War on Terror demonstrates how the force targeted Indigenous people and Indigenous-led groups, including Idle No More.[13] These studies are but a few examples that illuminate the ways in which surveillant policing operates to reproduce the control and logic of the settler colonial state of Canada, with the War on Terror serving as a formative logic for its ongoing justification.

The artworks in this chapter come together through a shared vision of the need to question policing in its relation to surveillance logic, a vision that reveals the contours and contestations of Canada as a settler colonial project. At the same time, each offers a different strategy to trouble wider understandings about why privacy is granted to some while disclosure is demanded from others, and about who gets to access and shape the personal information of the public and to what ends.[14] In contemplating these questions, the creative works position themselves and their audience in a complicated relationship that interrogates what surveillance scholars Haggerty, Dean Wilson, and Gavin J.D. Smith describe as "the ambiguities, paradoxes and often knotty interplays between the surveillers and the surveilled."[15] These works play with the frictions between seeing and being seen, enabling those who are normally objectified by surveillance capture to assert their subjectivity. In these works, the artists confront the question of who determines the context, interpretation, and control over policing narratives, or what information scientist and surveillance studies scholar Bryce Newell labels "the information politics of policing."[16]

The artworks here are organized around three main curatorial themes that consider different ways surveillance is operationalized within a logic of policing: profile, access, and technologies. In "Profile," the artworks illuminate predictive policing strategies that build criminal profiles as extensions of the biases of the surveillance watcher, while, in "Access," the works centre specific surveillant policing technologies to question the merits of the science – and attendant claims to neutrality – that policing agencies call on to publicly justify those strategies. In "Limits," the creative projects focus on surveillant policing technologies to illuminate the gaps in the knowledges generated by them. These thematic sections and the artworks included in them work together to create a larger and more robust sense of the interconnections between surveillance and policing in the contemporary Canadian context, and how creative works can contribute to understanding and challenging the role of surveillant policing as a characteristic of the settler colonial project.

Profile

The two artworks in this section share a strategy of taking familiar practices of criminal profiling used in surveillant policing and destabilizing them. They highlight bias and discrimination in criminal justice and law-enforcement systems – systems that rely on predictive policing and wide-scale data analysis to build criminal profiles – as a reflection of the larger structures of settler colonialism, in their intersections with capitalism, white supremacy, racism, cisheteropatriarchy, and ableism. These artworks demonstrate that surveillant policing and its use of predictive technologies replicate the biases that exist within Canadian settler colonial society, and they engage audiences to confront the ways in which surveillant policing models presume the criminality or guilt of those caught within their targeted frame. As sociologist Simon Egbert and criminologist Monique Mann write, predictive policing is a "socio-technical assemblage" that encompasses "not only the technical predictions themselves but also the enactment of the predictions on the street level by police – which can also have serious ramifications including discrimination."[17] These scholars affirm the need to "consider the broader socio-political and historical contexts from which these technologies emerge," to better understand the larger potential biases and issues predictive policing strategies can reproduce.[18] Following Egbert and Mann, I analyze the artworks in this section to think through the way in which surveillant policing creates criminalized profiles that replicate the beliefs and logic of the settler colonial system from which they emerge, and to provide insights into which bodies and communities are protected, and which are caught in the crosshairs.

Edwin Janzen's *Conspiracy Case: A Fantasy of Total Surveillance* (2012, fig. 2.1) features installation and performance elements and relies heavily on humour and caricature. Exhibited in July and August 2012 at the Station House Gallery in Williams Lake, British Columbia, *Conspiracy Case* involves Janzen creating his own satirical law-enforcement agency to investigate and surveil the activities of local burrowing animals, specifically targeting ground squirrels and marmots. Janzen transformed the gallery into a police-style headquarters from which his fictional law-enforcement agency – INTERPOL International Task Force on Rodent Crimes (IITFRC) – would conduct its investigative business. The headquarters includes a metal office desk scattered with work papers, pens, and empty coffee cups. Bulletin boards with area maps and notices of rodent crimes are mounted throughout the space, including on the wall above a smaller desk with a vintage-looking typewriter and a stack of paper. On the other side of the room, a small TV playing surveillance footage of the animals sits across from a large, free-standing bulletin board containing photos, text, and charts, a trope known by anyone familiar with detective and crime television genres. Janzen himself acknowledges his desire to replicate a recognizable popular culture version of a police investigation room with his installation: "The visual aesthetic of the police offices in 'The Wire' – the battered furniture, the bulletin boards overflowing with pinned-up data, the CRT [cathode-ray tube] monitors, coffee-making paraphernalia – is what I am after, here. An investigation must look like an investigation."[19] The installation and performance extends outside the gallery space. "Wanted" posters appear throughout the town in order to "invit[e] informants to come forward (many did)."[20] In creating a mock scenario of analog policing strategies used for surveillance, Janzen's project highlights structures and systems in place to map, track, and monitor the everyday activities of individuals.

As part of the surveillance activities in the piece, still and video cameras collect evidence of rodents' suspicious activities in the surrounding areas. As Janzen states, "Accepting the rodents' guilt as assumed, I surveil their activities using still and

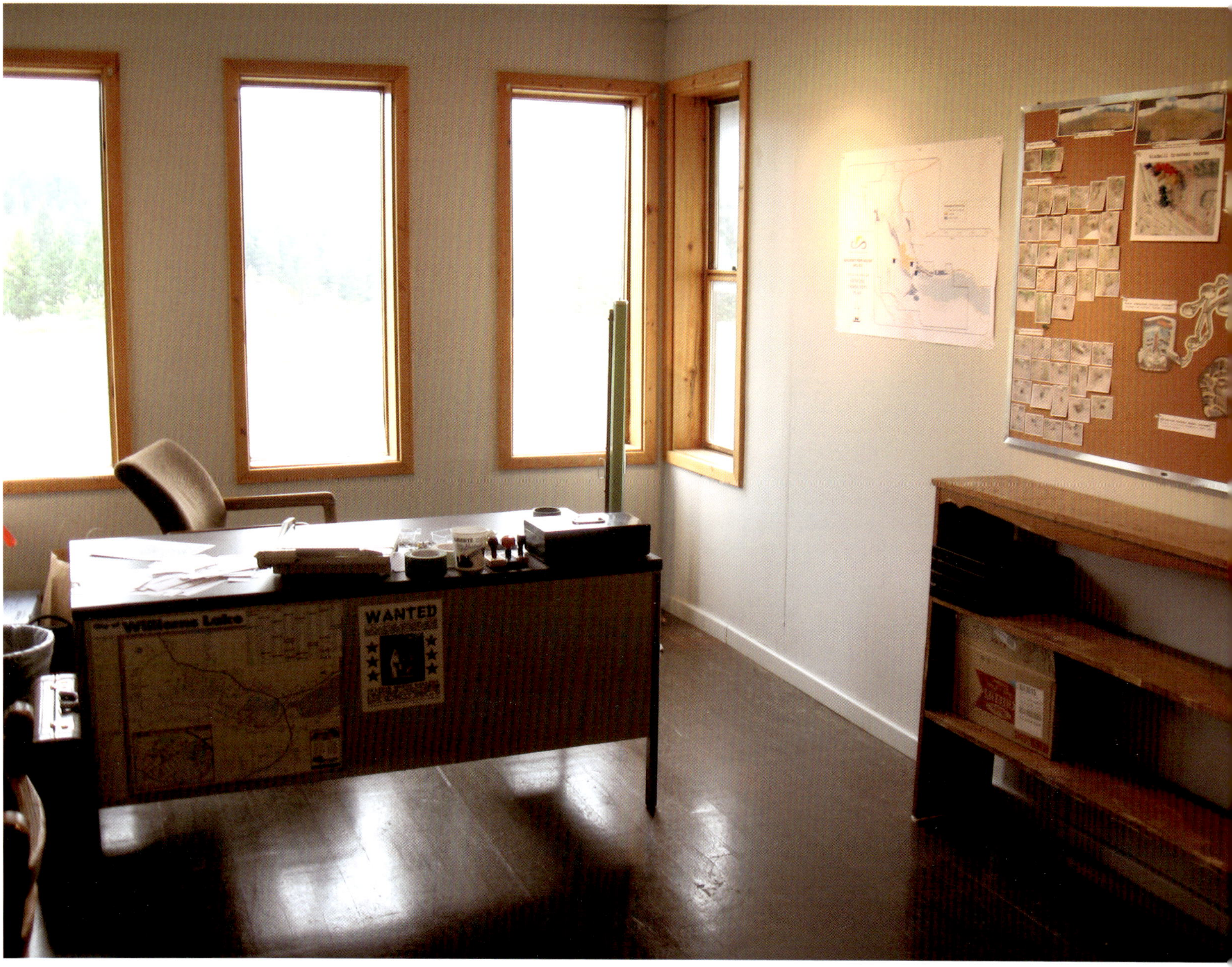
Williams Lake
WANTED

2.1 • Edwin Janzen, *Conspiracy Case: A Fantasy of Total Surveillance*, 2012.

video cameras, compiling police dossiers on them, identifying criminal relationships and networks, and mapping their tunnel complexes in a speculative fashion."[21] For the artist, the surveillance aspects of the project work to associate all those activities and animals captured with guilt. Surveillant policing, in this sense, functions to reveal and confirm crime and guilt; that is, to be caught within the web of surveillance is to be constituted as already guilty. The title of the artwork, *Conspiracy Case*, further emphasizes this phenomenon. In legal terms, conspiracy refers to when two or more people agree to commit an unlawful act, even if the act itself has not yet been committed.[22] In titling his artwork in this way, Janzen accentuates the subjective nature of assessing intent, and by extension, what is classified as criminal or unlawful. He emphasizes the extent to which criminality as determined through surveillant policing is evaluated based on expectations of what the assumed perpetrator *will* do, even if they have done nothing up to that point. While in Janzen's artwork the humour of it all is that the targets are rodents and other local wildlife under suspicion of being about to commit a "crime," the logic of this policing strategy has real and violent implications in the everyday lives of people.

Janzen uses humour to create the sense of discrepancy in his parody, which as art historian David Smith highlights, works as a creative strategy only when "there is a norm to contradict."[23] In this case, the norm this artwork contradicts is the common predictive policing tactics used by policing units across Canada. For example, the practice of "carding" – arbitrarily conducting street checks, whereby a police officer collects identifying information from civilians based on a subjective assessment that they might be suspicious – has long been criticized by citizen groups and humanitarian organizations for its inequitable targeting of Black people.[24] An Ontario Human Rights Commission report from 2023 found that this practice, whether official or not, is anchored in racial profiling and functions to funnel racialized people into the criminal justice system.[25] Studies across Canada have found that surveillant policing practices that aim to target suspicious individuals – rather than practices that aim to respond to unlawful acts – disproportionately single out racialized people and communities.[26] Journalist, author, and activist Desmond Cole chronicles a year in

Canada (2017), providing a month-by-month account of racist actions by police and other agencies that inequitably target and criminalize Black youth and communities.[27] Maynard's *Policing Black Lives* provides a parallel account with her historical overview of anti-Black institutional policing violence in Canada.[28] These studies, among others, illuminate the racist logic that drives and defines surveillant policing within the context of settler colonialism in Canada.

Janzen's parody of such systems in *Conspiracy Case* brings the logic of surveillant policing to light and problematizes it. The installation questions the assessment criteria for who gets targeted as a threat. Specifically, Janzen draws attention to the arbitrariness of proximity as a fundamental assessment criterion, given that targets are implicated simply by being in the area where police surveillance is taking place. While a mock law-enforcement agency targeting ground squirrels is the narrow topic of the work, I see the broader critique in *Conspiracy Case* as how predictive policing strategies that evaluate suspicion do so via racist, white supremacist values – along with other systems of violence and oppression – that make up the foundational ethos of Canada as a settler colonial nation. The humour, then, is a political strategy; it is a way to reveal and highlight some of the truths of surveillant and predictive policing models. By parodying this system, Janzen is able to use his artwork to reveal normalized truths about surveillant policing and the tactics – and biases – it deploys.

Derek Curry and Jennifer Gradecki's installation and interactive website *Crowd-Sourced Intelligence Agency* (2015, fig. 2.2) is less satirical than Janzen's *Conspiracy Case* and more focused on publicly replicating hidden systems. Installed from 17 October to 14 November 2015 in Calgary's The New Gallery as part of the group exhibition *Surveillance*, this interactive artwork asks participants to assume the role of a cyber intelligence analyst. In the gallery space, a computer keyboard and monitor allow visitors to access an online database and participate as security and policing agents by evaluating online posts. Participants are asked to review postings from social media forums such as Twitter. "Agent Bayes," an algorithm security system, flags potential threats from open source intelligence and marks certain Twitter postings

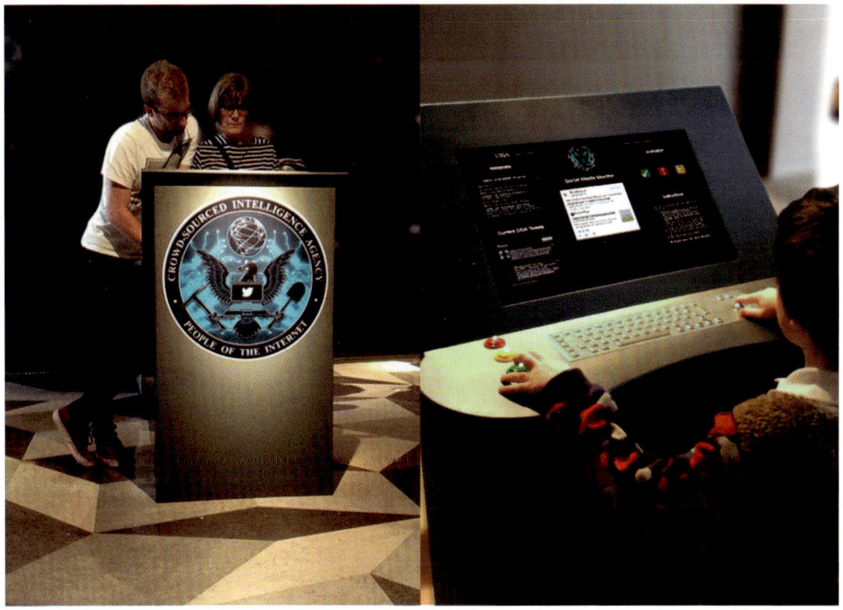

2.2 • Derek Curry and Jennifer Gradecki, *Crowd-Sourced Intelligence Agency*, 2015.

as "concerning."[29] These flagged posts, which are real, are then presented to the participant to evaluate as either neutral or problematic.

With each "concerning" tweet, the participant is provided with the online profile of the poster. After the participant reviews the data, they are asked to determine if the tweet is a valid cause for concern and to explain the reasoning behind their decision. The participant's review of the extensive list of flagged tweets makes clear that a substantial amount of superfluous data is not filtered out by Agent Bayes. Instead, the algorithm flags numerous social media posts that are unrelated and unremarkable, which betrays its vague and broad assessment criteria for what could be deemed "threatening." As Curry and Gradecki explain, *Crowd-Sourced Intelligence Agency* "replicates the data processing of an open source intelligence (OSINT) surveillance system ... By allowing users to experience how these systems frame social media posts and (mis)interpret natural language, especially slang, jokes, and sarcasm, we hope to provide a visceral heuristic of the process to help participants of our app to ask questions and make informed decisions about the large-scale monitoring of open source, social media data."[30]

The premise of this project is to highlight the way in which Big Data processes and manufactures systems to identify threats. *Crowd-Sourced Intelligence Agency* foregrounds how security organizations frame social media forums – including, but not limited to, Twitter – as influential platforms on which terrorist organizations share extremist beliefs and radicalize others, contributing to violence around the world. The increasing online presence of groups labelled as terrorists has become a growing concern for national security agencies, which cite this increase as justification for new surveillance methods to monitor and evaluate them. As a result, policing and intelligence agencies establish large-scale dataveillance practices intended to capture huge amounts of information, with the expectation that, as Curry and Gradecki state, "if enough data can be collected, future actions can be predicted with a high level of accuracy."[31] The problem that this artwork brings to light is that, during the machine-learning process, inaccuracy and bias gets baked into the algorithm, which then processes and scans data for predictive policing purposes. The scale – along with the inaccuracies and biases – of these policing projects is often hidden, or unremarked on in the public space. It is within this context that Curry and Gradecki situate their project, replicating these very structures while allowing the public to "crowd source" and analyze what the algorithm deems suspicious, thereby pulling back the curtain on the activities of intelligence agencies.

The project illuminates the inherent difficulties that cyber intelligence agencies face in developing a reliable system that can decipher what is harmless and what are real threats to national security. It reveals the level to which personal assessments and systemic biases infiltrate the process of determining what or who is problematic or threatening and why. Certainly, it provides space to contemplate the uneven policing of Black, Indigenous, racialized diasporic, 2SLGBTQIA+, and disabled people in Canada historically and today. The work also illuminates the social, administrative, and legal structures that enable those who occupy the privileged normative position within settler colonialism to function with presumed invisibility within surveillance systems, even as government agencies promote these systems as increasingly necessary to ensure national and global security. One example of this tendency for white settler bodies to escape the scrutiny of surveillance is the 29 January 2017 attack in Quebec City, when a white man walked into a mosque during prayer and opened fire, killing six people and injuring nineteen. The six victims who were killed in this shooting are Ibrahima Barry, Mamadou Tanou Barry, Khaled Belkacemi, Abdelkrim Hassane, Azzeddine Soufiane, and Aboubaker Thabti. The white murderer who shot and killed these men was known to many immigration and refugee lawyers and activists in the province because of his consistent trolling of online forums, where he spouted xenophobic, white nationalist views and asserted his support of US president Donald Trump and far-right French politician Marine Le Pen. Yet despite his online notoriety and consistent use of racist hate speech, the shooter was completely unknown to police, Communications and Security Establishment Canada, and the Canadian Security Intelligence Service in a moment when online surveillance of "anti-terror" operations was at an all-time high.

Such a major lapse in the surveillant policing and security apparatus that monitors online postings, which *Crowd-Sourced Intelligence Agency* replicates, can be set in contrast to the massive policing and surveillance operations that target the very communities that were victimized by this terrorist act, specifically, Black, Brown, and Muslim people. In her study of the experiences of Muslim youth growing up in the wake of the 9/11 attacks, sociologist and critical Muslim studies scholar Jasmin Zine critiques the ways in which Muslims "in Western nations … are subject to domestic security policies that label them as potential terrorists and enemies of the state."[32] Within the Canadian context, the surveillant policing apparatus is finely attuned to targeting Muslim communities, while the systemic threat of white supremacist violence remains largely outside its field of vision. This asymmetry reveals how surveillance funtions within settler colonialism: it treats whiteness as normative and non-threatening, thereby protecting white supremacy by rendering it invisible, even as it hyper-visibilizes racialized communities as objects of suspicion. The dominant biases of Canadian society are encoded in the surveillance algorithms and systems that police it. In doing so, the apparatus not only fails to prevent white supremacist violence but, significantly, actively sustains the conditions that allow it to flourish.

In creating a scenario that mimics the way in which personal information is surveilled, processed, and evaluated, *Crowd-Sourced Intelligence Agency* provides insight into how these surveillant policing policies are carried out, as well as the consequences of their conclusions. By providing firsthand experience for audience participants to confront and navigate these systems and their algorithms of suspicion, *Crowd-Sourced Intelligence Agency* exposes "potential problems, assumptions, or oversights inherent in current dataveillance processes in order to help people understand the effectiveness of OSINT processing and its impact on our privacy."[33] By replicating surveillant policing processes that proliferate globally en masse but often evade public scrutiny, this project provides participants with a creative platform to confront mass automated dataveillance and Big Data capture. In their engagement with the artwork, participants are asked to contemplate what and who is rendered visible or invisible, and how such distinctions are tethered to ongoing histories of settler colonialism in Canada that accept certain people and communities as either safe or a threat to the security of the nation-state.

Conspiracy Case and *Crowd-Sourced Intelligence Agency* employ a similar strategy: they take familiar practices of criminal profiling within surveillant policing and work to unsettle them. Both engage questions of bias and discrimination in criminal justice and law-enforcement systems, particularly those that depend on predictive policing and large-scale data analysis to construct criminal profiles, although each focuses on a different tactic of surveillant policing: *Conspiracy Case* relies on older, analog modes of police work – such as the charted bulletin board, the criminalized image of a mugshot, the surveillance footage on a screen, and public postings of "wanted" posters – that have long held meaning in popular culture. *Crowd-Sourced Intelligence Agency*, on the other hand, presents decidedly contemporary technological interventions by policing units that rely on social media, data capture, and artificial intelligence. Yet both demonstrate the ease with which a person (or a burrowing mammal, in *Conspiracy Case*) can be criminalized by surveillance systems that presume the automatic guilt of anyone caught within their frame. They also reveal the inaccuracies, assumptions, and oversights of these systems. These artworks highlight the discrimination of predictive policing strategies and how surveillant policing generates – or omits – criminalizing profiles that replicate wider social and political biases. Operating successfully through the model of art audience participation, both *Conspiracy Case* and *Crowd-Sourced Intelligence Agency* arrange a mode of engagement where audiences must confront their own presumptions, positionalities, and understandings of far-reaching issues associated with surveillant policing models and their strategies.

Access

This section builds on the discussion in the preceding section by focusing on three artworks that deploy different surveillant policing technologies – visual surveillance, facial recognition software, and forensic DNA phenotyping – to think through some of the ways in which surveillant policing enables access to people and their data, and how it uses this information to generate particular narratives and understandings of the world. Data accessed and gathered through surveillant policing technologies are often framed as neutral, impartial, and value-free, independent of human bias or social context. This framing positions technology as producing accurate, reliable outcomes based purely on logic and empirical evidence. Yet, this perception minimizes the structural inequalities embedded within the data collection and decision-making processes. The artworks in this section challenge what media and gender studies scholars Allison Page and Carmen Jones identify as the connection between strategies of police surveillance and "technological fetishism (especially the endowment of technology as an unqualified social good)" as well as the "assumption that technology is both highly efficient and 'neutral.'"[34] These creative projects explored here reveal and trouble these assumptions and show art audiences the inner workings of surveillant policing systems. In so doing, they demonstrate how surveillant policing technologies often reproduce and legitimize systemic inequities of racism, white supremacy, cisheteropatriarchy, and ableism within settler colonial regimes.

To challenge mainstream expectations of policing operations and the access they grant, the three artworks examined in this section offer, as communications scholar Roger Stahl and political scientist Sebastian Kaempf write, "alternative and counter performances ... to disrupt and dislodge official regimes of vision."[35] These works adapt surveillant policing tactics and present them to art audiences in new ways in order to draw attention to underrecognized institutional structures. In this, they work in conversation with Mariame Kaba's assertion that "surveillance is never neutral and that it is situationally weaponized."[36] The artworks reveal how the technologies used to access and interpret personal data build on pre-existing biases that replicate the dominant belief systems of settler colonialism and its applications of surveillant policing. These artworks challenge the idea that the implementation and interpretation of surveillant policing operations are neutral or unbiased. In this process, they ask: Who has access to my data? Through what means? And to what ends? Providing a creative way to grant their audiences access to technological systems of surveillant policing that are often obscured from public view, these artworks draw wider public attention to the inequities of surveillant policing and its potential violence within the Canadian context, particularly toward historically over-surveilled people.

Aljumaine Gayle's installation *Twenty_Four_Seven/365* (2019, fig. 2.3) uses surveillance cameras – the type often utilized by policing agencies – to highlight the disproportionate surveillance of Black communities in Canada. Produced in partnership with Black Artists' Networks in Dialogue (BAND) and displayed at the BAND Gallery in Toronto, Gayle's piece reveals implementations of surveillance systems by police agencies and governments. As Gayle articulates, *Twenty_Four_Seven/365* "explores state control and surveillance of communities of colour, specifically black and indigenous communities, through technological structures."[37] Surveillance cameras are mounted high on the wall of the gallery, intentionally positioned to be unseen and to simulate the familiar top-down camera angle of the surveillance gaze. As the

2.3 • Aljumaine Gayle, *Twenty_Four_Seven/365*, 2019.

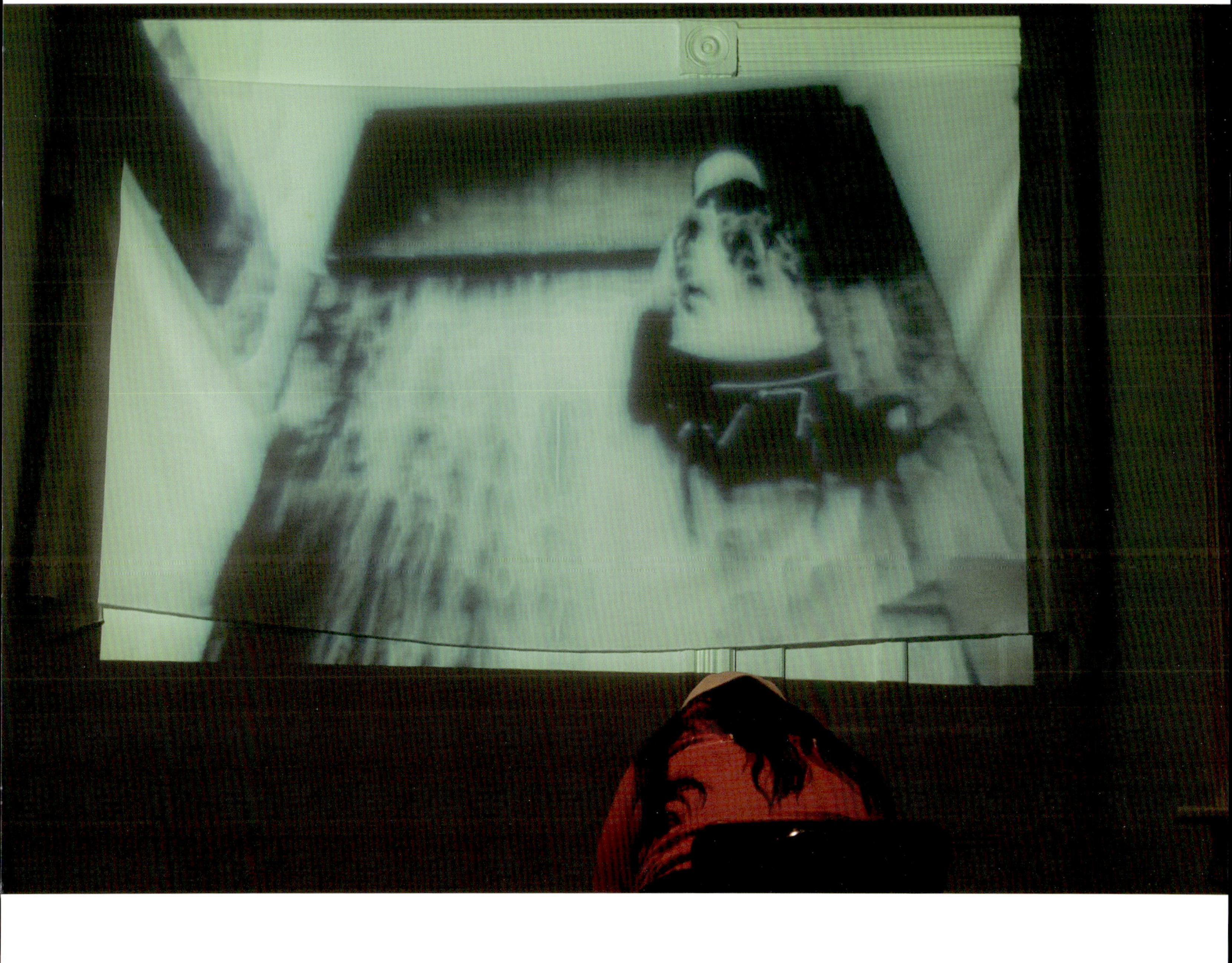

visitor moves through the space, their image is captured by these cameras, and they are confronted with their own life-size image projected on a gallery wall. The projected image is distorted by blurred outlines and colour trails, mirroring their bodily movement but contorting the finer details of their likeness. The resulting image, as presented to the viewer, both reveals and distorts.

Because of the camera's placement, oftentimes visitors are unaware that they are being recorded until they encounter their own distorted image. With this approach, the piece raises the issue of consent when it comes to surveillance. *Twenty_Four_Seven/365* compels exhibition attendees to subject themselves to surveillance – willingly or unwillingly – and encourages contemplation about how this process, far from being an egregious subversion, merely mimics the over-policing and over-surveilled experiences of many outside the parameters of the art space, particularly among Black, Indigenous, and racialized diasporic communities. In the public domain, the issue of consent is overridden by mainstream discourse of surveillant policing as a matter of public safety: the stated intention behind police surveillance is that it ensures safety and prevents crime within communities. Technologies such as ShotSpotter (which situates microphones that alert law enforcement about gunshots), facial recognition (an AI system that matches a face against a database of available facial profiles to identify people), biometrics (body measurements or calculations that are used to identify people), and Rekognition (an Amazon company that can be used for facial recognition) are promoted by private industry for public use in the name of preventing crime. Police in a number of districts are implementing more and more strategies of surveillance, purportedly for the same reason. For instance, police in Toronto, where Gayle's exhibition was mounted, use Clearview AI's controversial facial recognition software to identify potential suspects. Clearview's software matches images provided by the police with its database of over ten billion images collected from social media.[38] Yet multiple studies demonstrate that predictive policing measures through these types of surveillance operations do not deter or prevent crime; rather, they consistently target those communities that are most at risk of biased treatment and over-policing.[39]

In many ways, Gayle's *Twenty_Four_Seven/365* challenges claims that connect surveillant policing activities to crime prevention and public safety. Maynard articulates the realities of increased policing for Black people and communities in Canada, noting that "Black people in Canada are subject to invasive police surveillance that makes it difficult to exist in the public space."[40] In this context, surveillance tactics to ensure public freedom from harm sit in contradiction when the public is Black. Gayle purposely attends to these surveillant policing realities for Black people. In his artist statement, he quotes African and African diaspora and surveillance studies scholar Simone Browne's concept of racializing surveillance as "signal[ing] those moments when enactments of surveillance reify boundaries, borders, and bodies along racial lines, and where the outcome is often the discriminatory treatment of those who are negatively racialized by such surveillance."[41]

With his artwork, Gayle creates a microcosm where audiences can recognize themselves in these processes and ask themselves who they are through a surveillant policing lens. Gayle invites visitors to ask: Where and when might I be surveilled without my knowledge? How will my image register, and what implications and assumptions might be transposed via my visual markers? How will my actions and person be interpreted and understood within the context of surveillance and policing logic in Canada? In this way, *Twenty_Four_Seven/365* takes on issues of consent, interrogating racially motivated over-policing and the realities of surveillance capture as a tool for anti-Black racism. It highlights the realities of how surveillant policing activities deploy technologies to access and track racialized communities, a motivation that is circumscribed by racism and white supremacy as key tenets of settler colonialism. With *Twenty_Four_Seven/365*, Gayle presents the lived conditions for Black and other racialized communities in Canada and uses the technologies of surveillant policing that often oppress as a creative and critical tool with the intention to inform and resist.

If surveillant policing is conventionally framed as a mechanism of revealing and uncovering, then resistance toward it is often defined in terms of evading or hiding. The idea of concealing one's identity has been central to discussions about how to stop surveillant policing technologies from accessing and encroaching into people's private lives. Artists engaging with the potentials of rethinking surveillance often play with this idea, creating projects that imagine alternative ways of living under the surveillant policing eye. Unlike Gayle's piece, which speaks to the Black and racialized experience of surveillant policing, artist Leonardo Selvaggio invites audiences to play with surveillance as a white man in order to read the dominance of white supremacy within settler colonial surveillance logic against itself. Selvaggio's URME *Surveillance* (2014–ongoing, fig. 2.4) – pronounced "You are me Surveillance" – uses a strategy of masking and concealing to disrupt surveillant policing systems designed to reveal and make known.

Selvaggio, a Chicago-based artist, developed URME *Surveillance* to illuminate mass surveillance capture in public spaces and to offer everyday citizens a potential tool for resistance. Exhibited at the 2015 Art Souterrain Festival in Montreal, Selvaggio's work invites participants to wear masks in order to conceal their identities from surveillance cameras. Each prosthetic-like mask is an exact representation of Selvaggio's own face, 3-D printed on pigmented hard resin. The goal of this creative work is to flood surveillance captures with a single face – Selvaggio's own – thereby thwarting the identification and recording of individuals. Here, a single identity is multiplied into a collective one. The work is intended to serve as a defence against facial recognition surveillance software by linking the participant's actions to Selvaggio's identity; but it also obfuscates Selvaggio's own actions, because it simultaneously links his identity to multiple sightings in multiple places at once. URME *Surveillance* reframes surveillant policing structures that seek to record, capture, and identify individuals for the sake of building a digital identity that policing agencies may then target and criminalize.

This work blocks the surveillant policing desire to individuate and objectify; instead, as a strategy to counter the surveillance gaze, it insists on the group and the community, which is defined by donning the identity of one person en masse. The artwork illuminates a deep ambivalence behind surveillance logic: on the one hand, surveillance functions to identify and isolate singular individuals from larger

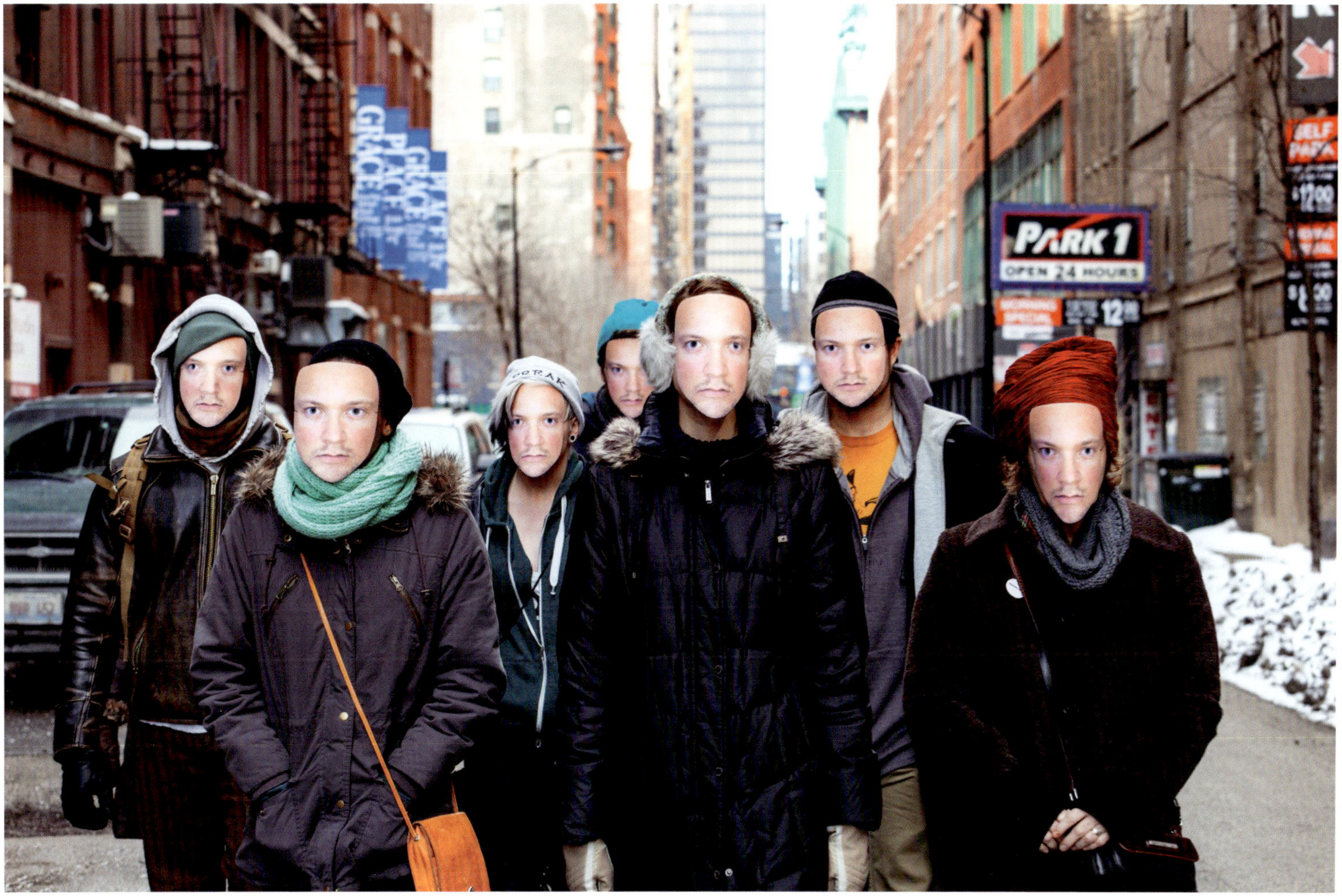

2.4 • Leonardo Selvaggio, *URME Surveillance*, 2014–ongoing.

groups, relegating those individuals to a category of suspicion, disconnected from larger communities of support; on the other hand, surveillance works to classify and assess individuals based on preconceived notions about their personhood and (at least perceived) membership in a group, a process communications scholar Oscar Gandy identifies as "the panoptic sort."[42] Political scientist Colin Bennett discusses this sorting as a form of categorical discrimination, whereby "individuals are arguably placed at risk because of their membership in, or assignment to, certain groups."[43] The feed from surveillance cameras throughout cities is run through facial recognition software designed to identify individuals and track their whereabouts. Thus, surveillant policing relies on its ability to assign identification – where identity is interpreted as finite, predictable, and determined – based on visible traits of the surveillant target. By offering a multiplication of his own image – that of a white man – as the surveillant target, Selvaggio refuses this logic of surveillance capture and, in turn, resists the unequal monitoring of certain bodies. The artist states that the installation "challenges viewers to consider the malleability of identities by misrepresenting and corrupting [their] own."[44] In so doing, Selvaggio's project enables the invisibility and unseeing of faces – both those behind the masks as well as his own, given its multiplicity – thereby impeding surveillance's identifying manoeuvres. URME *Surveillance* creatively presents multiple methods of using hypervisibilities and invisibilities as counterstrategies against surveillant policing forms.

URME *Surveillance* thinks through the relationship between surveillance visibility and invisibility, and the extensive networks of surveillant policing intended to target and record the actions of specific individuals. As Monahan describes, "masking of identity" enables participants "to undermine technological efforts to fix someone as a unique entity apart from the crowd."[45] Yet this masking also raises broader questions about what can be concealed. Selvaggio's use of his own white, male-presenting face as the model of uniformity demonstrates the privilege and politics of invisibility this position offers within the Canadian settler colonial context and beyond. It reveals which bodies are often considered indistinct, unremarkable, and unthreatening to surveillance capture. But it also highlights the ways in which the face is not the only marker used in surveillance identification. What happens, for instance, when skin not covered by the mask fails to match the metrics of the face? Does this process then hide or reveal? Moreover, although it is quite prominent, facial recognition is not the only metric used in surveillant policing identification; as Monahan writes, "there are many circulating, overlapping, and reinforcing markers of identity beyond the face, so the notion of evading them entirely is always a losing game."[46] Perhaps, for URME *Surveillance*, the goal is not necessarily to "evade them entirely," but rather to complicate and problematize the way in which these surveillant policing systems operate.

The work functions not because visitors are able to look at and contemplate the piece, but because they themselves become the diffuse art objects out in the world. Outside the space of the art exhibition, Selvaggio has made the masks available for online purchase, encouraging others to join in an "organized artistic intervention" against surveillant policing. In this way, Selvaggio provides people with the practical tools to navigate public spaces while deterring recognition by surveillance technology. Certainly, this artwork functions to confront audiences with the reality of mass surveillance capture, to question their own presumed visibility/invisibility within this policing process, and to invite them to think about how their identity and positionality impacts how they are served or harmed by these operations of surveillant policing. URME *Surveillance* illuminates the racialized and gendered biases – among other systemic inequities – of what bodies function within the Canadian state as unmarked norms. It questions how surveillant policing technologies respond not just to a person's actions but also, and more significantly, to assumptions about their identity – assumptions shaped by pre-existing hierarchies rooted in systems of dominance and oppression that hold power in Canada.

While Gayle and Selvaggio interrogate surveillant policing technologies through their access to visual or superficial capture, artist Heather Dewey-Hagborg reveals the micro level of surveillance technologies. Dewey-Hagborg, who is based in New York, received international recognition for her work *Stranger Visions* (2012–13, fig. 2.5). *Stranger Visions* includes life-size faces of individuals, generated in colour on a 3-D printer. The appearance of each face was approximated from DNA extracted from discarded hairs, cigarette butts, and chewing gum found on the streets of New York City: each likeness was computationally generated, derived by analyzing the collected DNA. This work was the centre point for a workshop Dewey-Hagborg led at the art studio/gallery Ada X in Montreal in 2014, which enabled audience members to see images of the work and its installations and to engage with the technologies that generate these types of objects.[47] In installations of *Stranger Visions*, the life-size faces are mounted on the gallery wall higher than the average height of a person, and are angled downward so they appear to be staring back at the gallery visitors. Each portrait is mounted above a box that contains information about the DNA profile and the site where the DNA material was found, along with a photo of that site. For Dewey-Hagborg, the connection between the printed portraits and the

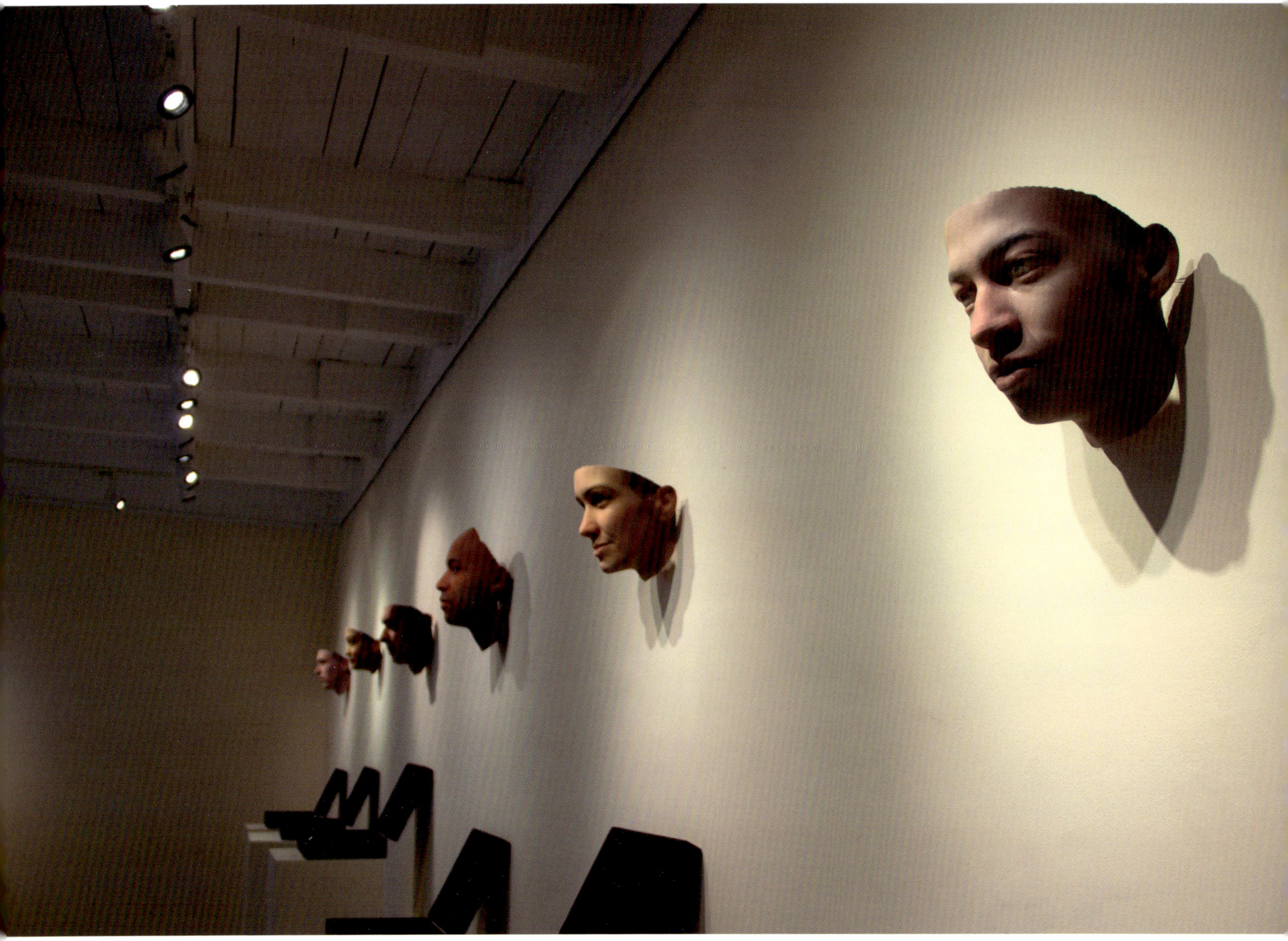

2.5 • Heather Dewey-Hagborg, *Stranger Visions*, 2012–13.

DNA materials that generated them is essential to the artwork: "Working with the traces strangers unwittingly left behind, the project was meant to call attention to the developing technology of forensic DNA phenotyping, the potential for a culture of biological surveillance, and the impulse towards genetic determinism."[48] The work recalls historical forms of artmaking such as sculptured busts but brings these aesthetics into the digital age, presenting new arenas of "biological surveillance" that track and monitor people and their bodies through discarded and forgotten evidence of their presence.

While this work feels like it belongs to the realm of science fiction, *Stranger Visions* actually illuminates current – or anticipated – models of surveillant policing technologies across Canada. For example, in 2014, the same year Dewey-Hagborg held her workshop in Montreal, the Toronto Police Service announced a partnership with two American companies, AKESOgen, a genomics company, and Identitas, which specializes in DNA profiling, to identify perpetrators in the city's cold case files. Detective Sergeant Brian Borg, the leader of the team tasked with addressing Toronto's cold cases, details the intent behind these collaborations: "to take DNA swabs from old crime scene evidence and turn them into police sketches of possible perpetrators."[49] This strategy of forensic DNA phenotyping involves, as sociologist Filipa Queirós explains, "a set of techniques that aim, from biological samples, to infer probabilistic information about externally visible characteristics of criminal suspects, such as eye, skin and hair colour, and also ethnicity, through the inference of biogeographic ancestry."[50] According to media reporting on these collaborations, the result is a "high-tech composite sketch."[51]

Coverage of DNA phenotyping as a policing strategy appeared in the *New York Times* in February 2015, with the article highlighting that law-enforcement units across North America – as well as the private companies providing the phenotyping services – were optimistic about the potentially enhanced profiling practices the technology purportedly offers.[52] While it seems that, to date, forensic DNA phenotyping as a policing tool has had limited success, this has not stopped Canadian police services from using it and releasing the resultant images to the public. Calgary police constructed an image that was circulated to the public for help with a 2017 case, and the Saskatoon Police Service considered forensic DNA phenotyping to assist a 2020 investigation.[53] Neither of these cases was solved using this strategy. In 2022, Edmonton police were forced to issue a public apology after circulating a sketch of a suspect based on forensic DNA phenotyping after a mass public backlash questioned its accuracy and criticized its contribution to further stigmatizing racialized communities.[54]

Stranger Visions reveals to audiences forensic DNA phenotyping as a form of biological surveillance. Rather than a neutral statement on the possibilities of these technologies, Dewey-Hagborg's piece raises essential questions about the validity of the technology's promises and the issues involved in its implementation. The way in which such technologies have already been utilized within the Canadian context illuminates that this science is not a neutral inquiry but rather is reflective of the larger biases prevalent in the settler colonial context. As Dewey-Hagborg explains in a published piece detailing her artwork, the logic of forensic DNA phenotyping is grounded in discredited pseudoscience of the nineteenth century, such as physiognomy, phrenology, and anthropometry: "forensic DNA Phenotyping is simply the latest in a long succession of identity-inscribing technologies which claim to use science to classify types of bodies into socially constructed categories like gender and race."[55] Rather than a conclusive representation pulled from the strands of DNA, forensic DNA phenotyping creates an image based on statistical predictions tied to the expectations of what certain genomes will most likely represent. So, as opposed to supplying a specific portrait of an individual suspect, for instance, these technologies produce images that are more accurately described as composites of the most likely characteristics associated with a broad genetic profile. Yet, when they are pulled together as a portrait and adopted by law-enforcement agencies – or by an art project bringing these behind-the-scenes investigative tools to the public – the process lends the image, as anthropologists Roos Hopman and Amade M'charek note, "an air of exactitude."[56]

The reality that forensic DNA phenotyping lacks specificity, juxtaposed with the possibilities of its use as a serious investigative tool by police, raises numerous ethical questions regarding privacy and use of evidence. *Stranger Visions* points to the danger that policing agencies might rely on assumptions and stereotypes to fill in the blanks left by DNA phenotyping. This work brings to the fore the ways science and biology may be called on once again to affirm social perceptions, particularly those bound to biological determinism. Certainly, as Dewey-Hagborg herself states, the goal of this piece is to trace this technology's journey from the realm of science fiction, to scientific research, to law-enforcement strategies, and "to create a public dialogue concerning the emergent possibility of genetic surveillance," not as a speculative future but as a concerning present.[57]

In the Canadian context, the technologies Dewey-Hagborg uses to generate her artworks are being pushed as surveillant policing strategies, while the ambiguity of the science behind them is not often made clear to the public. *Stranger Visions* confronts audiences with the disconcerting idea that their DNA could be plucked and turned into something considered definitive and authoritative. The way Dewey-Hagborg connects the discarded objects with retrieved DNA from public spaces to the corresponding portraits illuminates the ease with which biological surveillance can take place. "The possibility of reconstructing absent bodies based on DNA," as artist Élène Tremblay writes in a review of the work, "stresses the impossibility of eluding surveillance systems and remaining invisible."[58] By confronting audiences with the reality of these technologies, *Stranger Visions* raises awareness of this new form of surveillant policing capture and how its applications rely on social interpretations of composite phenotypes while also asserting biology as fact. These definitions are anchored in stereotypical assumptions of social categories that subsume forensic DNA phenotyping into the logics of settler colonialism.

The three works in this section shine a light on different surveillant policing technologies as a way to engage audiences and inform them about how their information might be accessed and interpreted by often opaque forms of policing and surveillance capture. In so doing, they complicate mainstream perceptions that justify surveillant policing: that these technologies are neutral and that they protect the general public. The surveillant policing operations these artworks play with carry the potential to extend policing's reach while maintaining historical and ongoing inequities. Each of these works, using its own approach and strategy, emphasizes which people and communities are most vulnerable and raises awareness around the surveillant policing systems that target them. They emphasize that these technologies extend the access granted to policing agencies while reinforcing pre-existing biases maintained by settler colonialism. These three artworks invite audiences to grapple with the specifics of how surveillant technologies work and the ways in which policing agencies wield them, as well as to notice the existing prejudices that are built into surveillant policing operations. The works demonstrate that surveillant policing operations are never neutral tools; rather, they bend to the underlying logics and biases of those creating and utilizing them.

Limits

The three works explored in this section focus on the limits of surveillant policing, highlighting the gaps and knowledges left uncovered by policing technologies. By "technologies," I do not necessarily mean computers, engineering, algorithms, or machinery in general; rather, I mean a set of surveillance devices that extend the capacity of policing agents to literally *see* what is going on. The creative projects in this section all engage with surveillant policing and tracking technologies that interpret meaning around bodies and their movement, specifically in relation to how forms of technology extend policing's reach. Like those in the previous section, the artworks here look at forms of surveillant policing technologies and interrogate what they reveal, who they target, and why; in addition, however, these works assess what they hide.

Technologies that enhance the ability to see, such as X-ray technologies, quantify and make material the lens of the surveillant policing gaze and aid in strengthening policing capacities – or at least give that impression. But other forms of surveillance technologies – ID cards, for instance – also participate in enhancing surveillant policing insights into private lives. The role of these technologies is to expose what is hidden and to grant elevated access into the private lives of people. As political scientist David Barnard-Wills and criminologist Helen Wells write, a major factor in the development and implementation of technologies designed for surveillance policing is the intention to render "previously private, or publically [*sic*] unproblematic behaviours ... visible and problematized, be they behaviours that mark the observed as risky or as at risk."[59] What is judged as problematic or criminal is always subject to interpretations shaped by the dominant social, political, and cultural orders that influence who is seen as a threat and who is not. As surveillance studies scholar David Lyon explains, "technological developments and social processes mutually influence and shape, or co-construct, each other."[60] The contemporary period of surveillance within the Canadian context is inextricably linked to longer histories of inequitable policing and the ways in which surveillant policing strategies work in tandem with the creation and maintenance of the settler colonial Canadian project. The artworks discussed in this section reveal the limits to the technologies of surveillant policing, both in what information they are able to gather and how that information is interpreted and understood.

Artist David Spriggs explores the technological and physical capacities of surveillant policing operations in his piece *The Visible Spectrum* (2015, fig. 2.6). This work premiered at Arsenal Contemporary Art Gallery in Montreal, for Spriggs's solo exhibition *David Spriggs: PRISM* in 2015. *The Visible Spectrum* is an installation of a video animation. The animation depicts a semi-truck carrying more than a dozen people in its trailer along with what looks to be stacked pallets of bottles and other materials near the trailer door. The images are computerized and made to replicate the look of X-ray vision, although a more detailed and colourful version than currently possible in the real world. Inanimate objects (such as the structure and mechanics of the truck along with the materials on the pallets) are illuminated in cool tones of blue, while the human bodies of the truck driver and those standing and sitting in the trailer are coded in tones of red, yellow, and green. As the video progresses through its four-minute loop, the colours wave and shift, lending a sense of movement, in

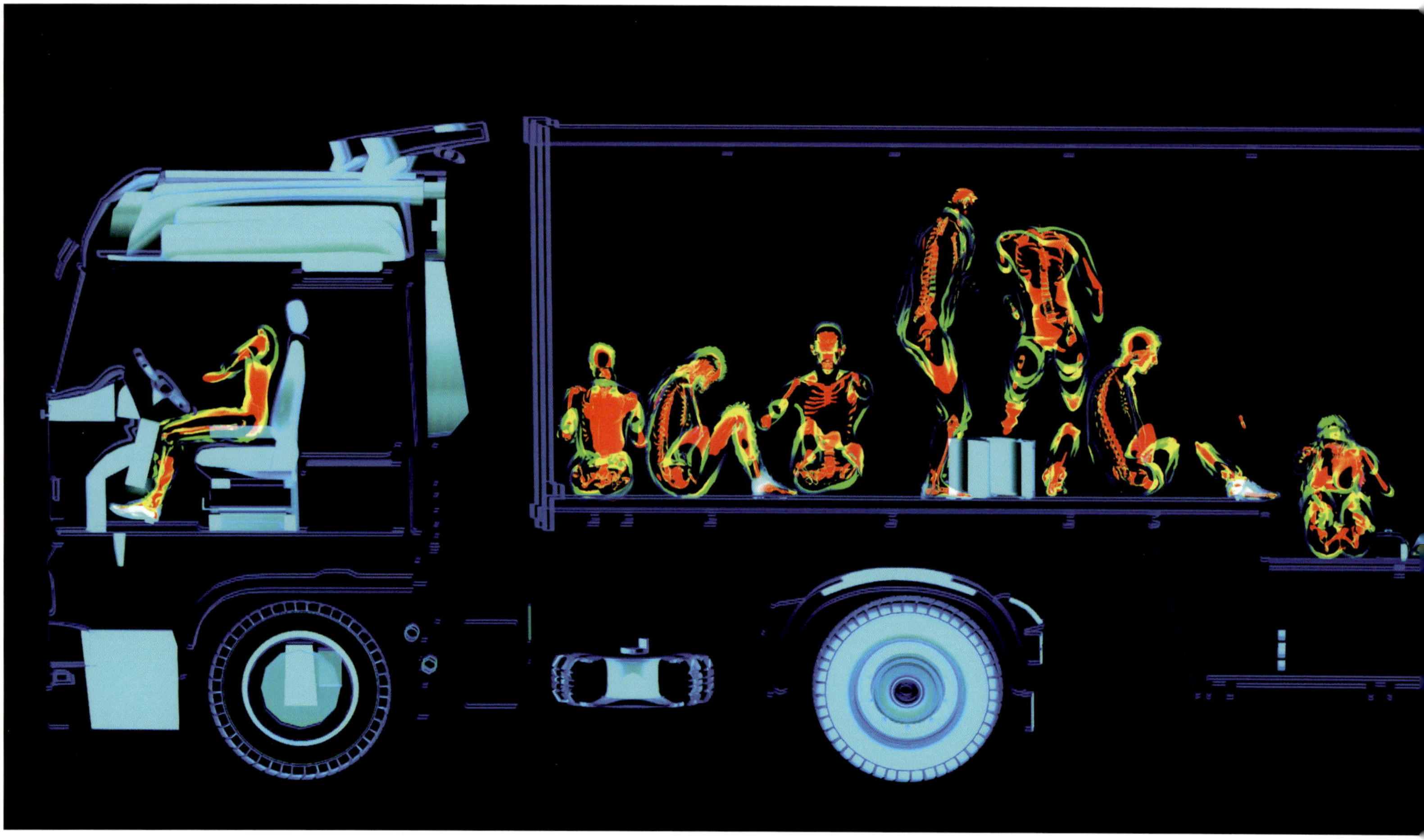

2.6 • David Spriggs, *The Visible Spectrum*, 2015.

terms of both the objects within the piece and the scanning technology itself. What is illuminated or in darkness shifts and changes throughout the loop, a constant rhythmic interplay between visible and invisible that never allows the full picture to be in focus at any point. The title of this artwork, *The Visible Spectrum*, refers to the electromagnetic waves that are capable of being detected by the human eye and highlights the enhancements of these capabilities through the use of surveillance technologies. In this piece, Spriggs brings forth questions of looking and access that surveillant policing technology grants, and interrogates what is illuminated in these processes.

To me, the content of the piece – people tucked inside a semi-truck trailer – gestures to contemporary issues of border controls, immigration, migrant smuggling, and surveillance technologies that monitor and police national boundaries. Bioinformatic forms of surveillant policing such as X-raying vehicles passing through national borders are not uncommon in the Canadian and US contexts. Since 2001,

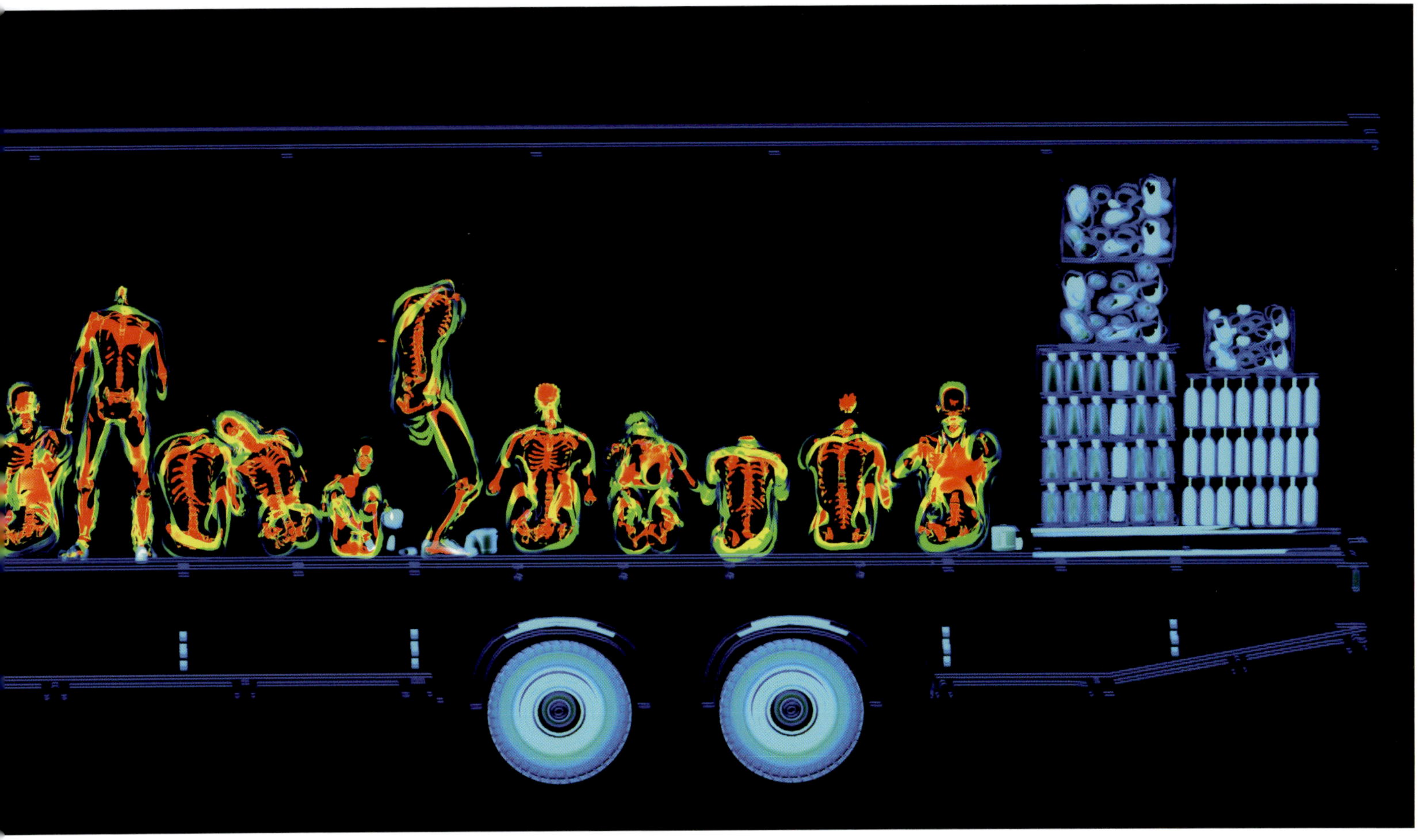

border patrols in both jurisdictions have increased their technological capacity to gain access into areas previously unseen through standard checks. Different forms of X-rays – such as HCVM (Heimann Cargo Vision Mobile) X-ray scanners, drive-through X-ray portals, and backscatter technologies (an advanced imaging technology that can provide more detailed images than conventional X-rays) – have become standard tools of national border controls. Yet, while the technologies are new, the border as a contested site of access and denial has a long-standing history informed by settler colonial and imperial ideals of who is positioned as internal or external to the nation-state. Communications scholars Tamara Vukov and Mimi Sheller note that national border spaces "are key sites of securitization not only because they are associated with drug trafficking and illegal immigration, but also because they constitute a global racialized divide based in colonial histories of population management, spatial differentiation, and bio-politics."[61]

Increasing xenophobia and immigration restrictions since 2001 have turned public fear and attention to the supposed dangers of illegal immigration. News media covering this issue often include images generated by surveillance technologies, including X-ray images of people crossing borders in the backs of trucks, particularly in the border regions of the southwestern United States. These are the surveillant policing technologies and cultural contexts I see Spriggs attending to in this piece, although he brings a new visualization to them. *The Visible Spectrum* introduces movement and life into the X-rayed image, reinvigorating the human figures with colour and movement, resulting in a hypnotic dance. This artwork is thus not just about the surveillance technologies of tracking and policing; it also recognizes the people themselves. It transforms the X-rayed representation of flat, whitewashed images of undocumented migrants into colourful and lively depictions of people on the move, a strategy that works to rehumanize the border encounter and the otherwise unfeeling manner in which surveillant policing technologies seek to capture and monitor mobilities.

The Visible Spectrum highlights those who are rendered hypervisible – and thus often categorized as intruders into the national space – by enhanced surveillant policing technologies such as border X-rays, and seeks to discover what is rendered invisible in the process. By giving life, colour, and movement to the people in the back of the truck – those often positioned as dangerous, unwanted intruders by these technologies – this artwork recognizes their humanity. It renders these bodies as people with lives, rather than as flat, ghostly spectres presented in official X-ray images. Spriggs's work attends to the increased technological remediation of border sites as police surveillance technologies strive to increase visibilities in these spaces. *The Visible Spectrum* then asks the viewer to question the relationship between that which is visible and that which remains unseen, and to notice the bodies and lives of the real people who get caught up in this process. This piece asks its audience to think through the conditions under which certain people are considered intruders, while others are defined as citizens, and how the border – and the surveillant technologies that police it – operates to preserve and expand such categorizations to affirm Canadian state claims to the land.

Another work that similarly situates representations of bodies at the centre of surveillant policing technologies at border sites is Kate McQuillen's *The X-Ray Series* (2013). This series, which reinterprets recognizable angles of vision and access of surveillance technologies to question their role in policing the body and its movements, was part of McQuillen's solo exhibition entitled *Backscatter (an exploration of surveillance in the wake of 9-11)*, held at the O'Born Contemporary Gallery in Toronto in 2013. *The X-Ray Series* comprises monoprints, reminiscent of security body scans, of various feminine undergarments concealing what appear to be weapons. The X-rayed image in her monoprint *Boxcutter III* (fig. 2.7) immediately recalls airport security, a particularly fraught place, where personal privacies, targeted suspicion, and terror-driven security surveillance intersect. McQuillen created these works in response to her feeling that she – or anyone – could become a target of suspicion at any moment: "I wanted to produce a body of work about my feelings concerning the extent of being under surveillance. I used my underwear to show the intimacy we're having with the government ... They are seeing more of me than I want."[62] *The X-Ray Series* questions elevated forms of post-2001 surveillant policing at border sites in terms of inclusion, access, threat, and criminality, and how this policing reveals the rationale of the settler colonial project.

The sense of ubiquitous government and corporate oversight certainly intensified in the post-9/11 North American environment, along with more extensive public discussion on security, privacy, and terror. *The X-Ray Series* works here to generate a feminist and gendered critique of bodies of desire in contrast with bodies of threat. For McQuillen, surveillance tactics, like X-rays, that afford access to private lives are invasive: indeed, she sees them as assaultive. Capitalizing on stereotypical and patriarchal associations between feminine bodies and sex, McQuillen uses the outlines of feminine lingerie to lend the project a sexualized and illicit effect. In so doing, she sets up a power imbalance between the watcher and watched, invoking parallels with predatory watching. As well, the unexpectedness of the soft, lacy fabric juxtaposed with the sharp, presumably dangerous metal prompts the audience to reflect on why this relation feels surprising and ask: Who would I normally expect to be carrying a

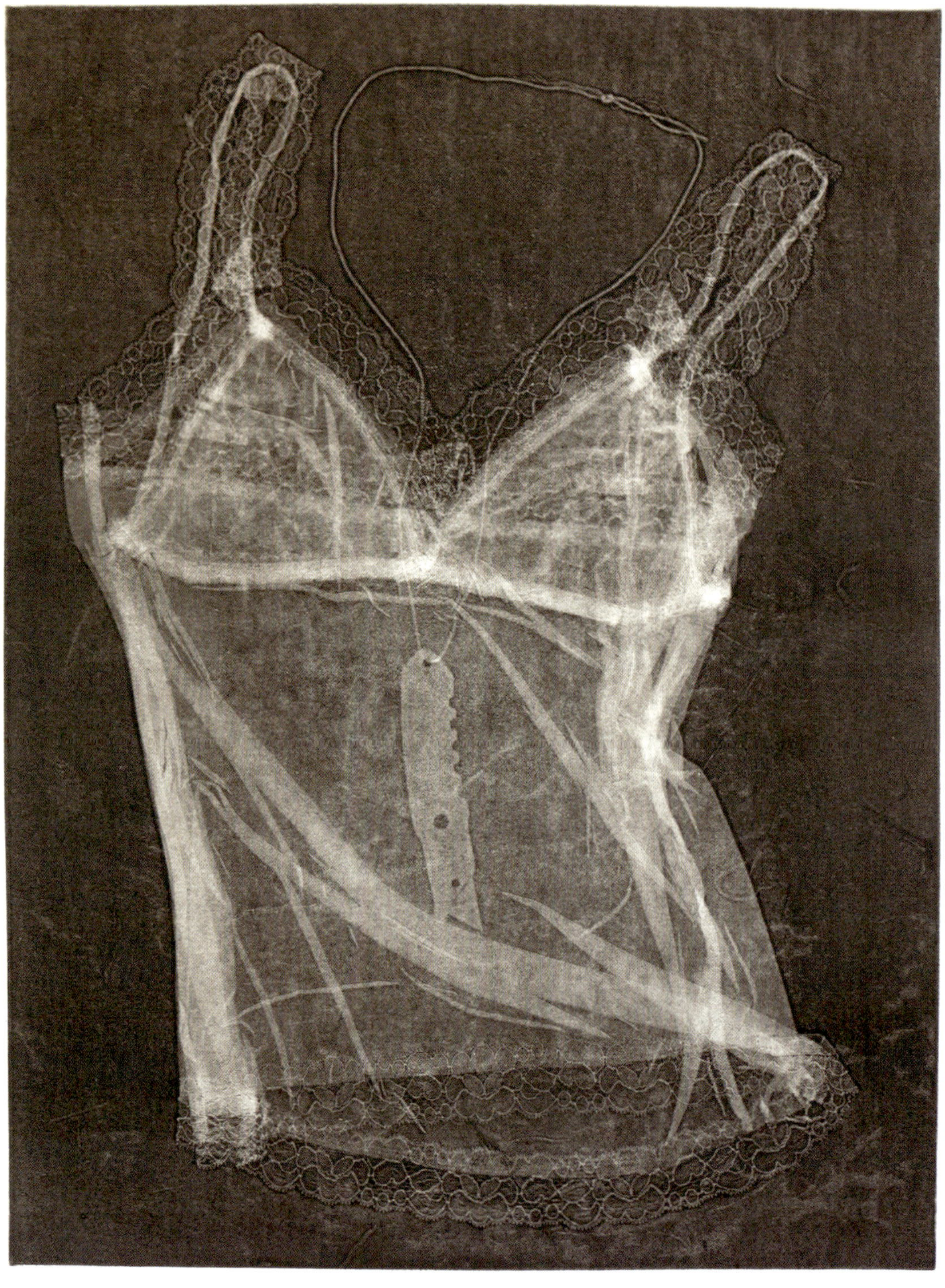

2.7 • Kate McQuillen, *Boxcutter III*, 2013.

weapon? What assumptions are my expectations based on? And what level of access do I assume is required to deem someone innocent or guilty? Caught in the lens of the X-ray machine, the juxtaposition between materials is heightened, projecting an element of suspicion, danger, or threat onto the image. However, despite their appearance on the X-ray, the threatening shapes are in fact harmless paper cut-outs.

To make the feminine lingerie sinister or threatening, McQuillen introduces the knife-shape, albeit one that is materially innocuous because of its paper composition. By implanting this apparently threatening object, McQuillen highlights what it means to introduce threat or uncertainty, in contrast to those bodies or images that are already presumed to be unsafe by the metrics of the surveillant policing apparatus in the post-2001 environment. Being categorized as potentially dangerous or unwanted through increased surveillance targeting has existed long before 2001 and the War on Terror, especially for Black, Indigenous, and racialized diasporic communities, as well as 2SLGBTQIA+ and disabled people. And certainly today, these people and communities are still most inequitably targeted by multiple forms of surveillance, not least of which include X-rays and surveillance measures found globally in airports in the post-9/11 environment. This post-9/11 targeting of certain people and communities affirms and expands the rationale of the settler colonial project, which works in tandem with other systems of oppression and inequity.

In relation to the use of surveillance technologies as a tool of white supremacy and racism, Simone Browne explains that policing technologies are an ongoing factor "in how racism and antiblackness undergird and sustain the intersecting surveillances of our present order."[63] I understand Browne to mean that the application of contemporary surveillant policing technologies is an extension of world views that are already tethered to racism, as part of a network of oppressive systems and ideologies. In her creative work, McQuillen gestures to the wide net that airport surveillance casts, forcing all visitors to expose themselves and their private lives to the eye of security and its technologies; however, she also highlights the inequity that continues to exist around who or what is rendered dangerous, and who or what is presumed threatening. *The X-Rays Series* emphasizes this large-scale information-gathering system and the threat of the all-consuming surveillance net, an interest inspired by Edward Snowden's revelations about previously undisclosed widespread monitoring by the US National Security Agency. With this work, McQuillen addresses how security procedures such as body scans can yield misleading results and leave innocent travellers feeling like suspects. So too, this work demonstrates how the recognizable surveillance lens – which renders invisible elements visible in the case of the X-ray machine – immediately produces a criminalizing, suspicious bent to any interpretation of who or what is caught in its gaze.

Technologies of surveillance are not always digital or highly technical. The modes and structures of surveillance manifest in a multitude of often seemingly benign, analog ways. While technologies of increased visibility, like the X-ray systems explored in *The Visible Spectrum* and *The X-Ray Series* provide increased access and insight into people's private lives, more mundane and even unexpected forms of data acquisition and tracking, such as identification cards, play an integral role in surveillant policing structures in everyday life. Included as part of the group exhibition *Sorting Daemons: Art, Surveillance, and Regimes of Control* held in 2010 at the Agnes Etherington Art Centre in Kingston, Ontario, Dave Kemp's *Data Collection* (2009, fig. 2.8) is composed of photographs of identification cards from 100 different people. Each person's collection of cards is arranged on a grey rectangle, which are installed in a large grid pattern on the wall. The photographs of the cards, which are the same size as the cards themselves, include bank and credit cards, school IDs, driver's licences, gym memberships, transit passes, and other cards that offer a glimpse into each person's habits, interests, communities, groups, and affiliations.

The identification information outlined on each card is available to viewers, although the participants who volunteered their IDs were allowed to remove any cards that they were uncomfortable with displaying publicly; these removed cards are replaced with a black square with the word "withheld" printed in white. In this way, *Data Collection* makes explicit those instances when private information has been removed from public view. Kemp's piece examines the circulation of information related to identity, privacy, and security, and the extent to which individuals willingly

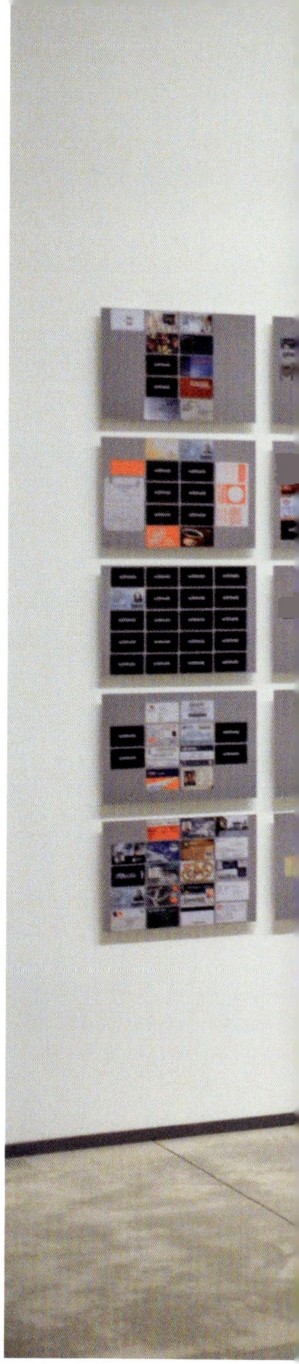

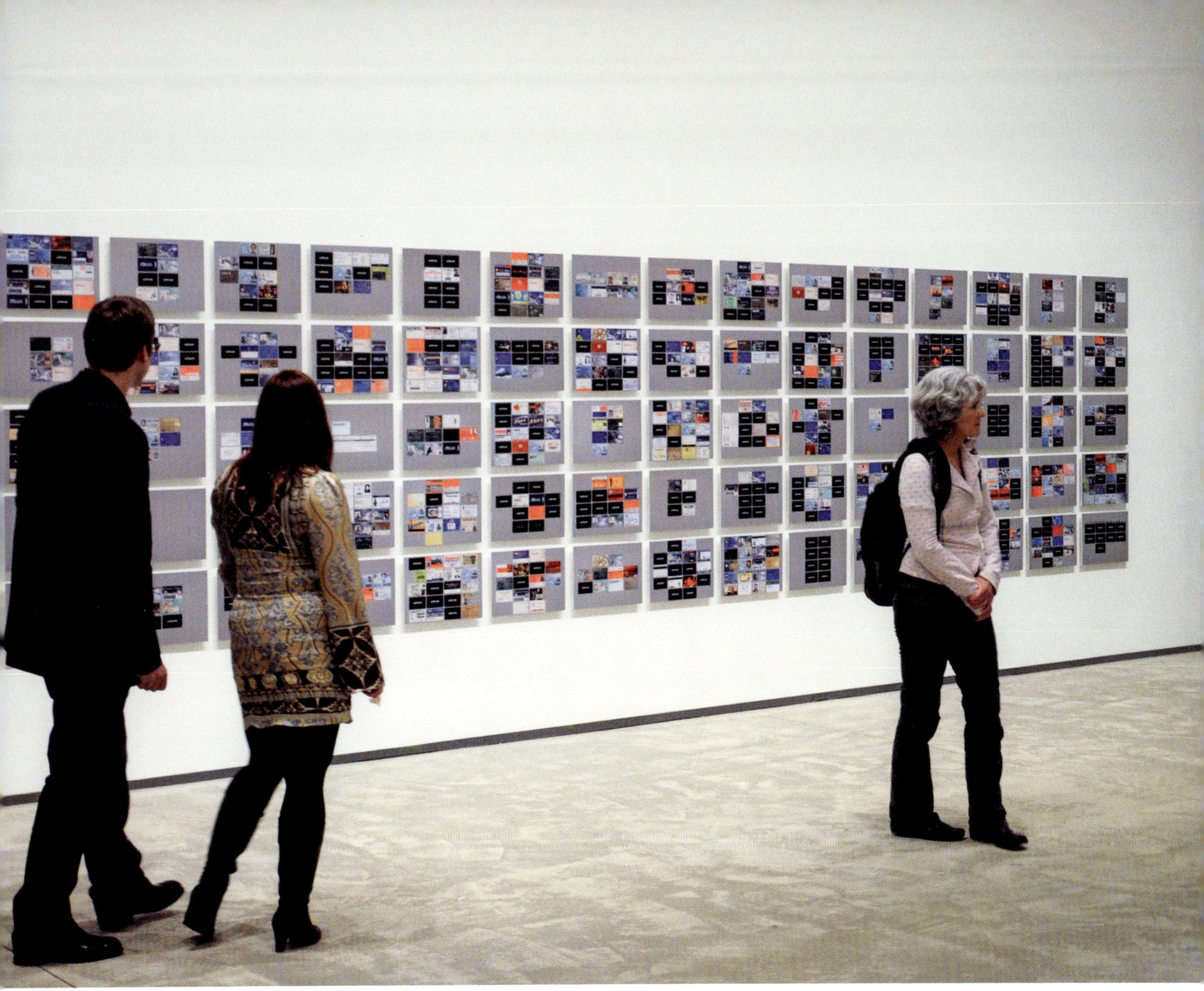

2.8 • Dave Kemp, *Data Collection*, 2009.

– if not intentionally – participate in offering their personal data to the surveillant policing apparatus. It questions who has access to that information and how it is being used and interpreted. The artwork also invites the viewer to reflect on the level of access – or denial – that they are being granted into each person's life and how this information functions as a representation of that life itself.

In many ways, each grey rectangle becomes a portrait of the card holder, one that characterizes personhood or individuality through the identification generated through the cards. The work dialogues with philosopher Irma van der Ploeg's assertion that, as "our bodies gradually becom[e] entities consisting of information – the body as data – the boundary between the body *itself* and information *about* that body cannot be taken for granted anymore."[64] *Data Collection* examines the space between identity and the data representation of identity: How much of our lives are presented through the access we grant governments and corporations? And how does that information define who we are? Certainly, it brings to mind David Lyon's symbols of "the stable self" – documents that function to prove our identities as well as assert our connections to the surveillance apparatus that allows or denies our movement and defines our very existence vis-à-vis the places we go.[65] In this piece, the personal data presented ostensibly demonstrate each person's capacity to exist and move through certain spaces: each card is a marker of that person's larger profile of how and where they move throughout their life. Yet, having an ID card does not necessarily guarantee access: as the discussion of "carding" in relation to Janzen's piece earlier in this chapter demonstrates, the surveillant policing apparatus interprets identifying documents differently depending on whether the person in possession of them is judged to be innocuous or criminal. Such assessments are predicated on judgments about the body that are inextricably tethered to the conditions of settler colonialism and how surveillant policing is part of a logic that determines movement and life on the land.

In presenting these details publicly as an art project, *Data Collection* positions the art visitor as the policing or surveilling agent, the one with access to the private lives of others. The installation of the cards as a large-scale grid on the gallery wall forces a particular type of engagement, whereby the movement of the visitor's body becomes part of the work's interpretation. To take in the entire piece, the viewer must step back, but to see the specifics of information on a single card, the viewer must come close. In this way, *Data Collection* creates a unique interplay between intimacy and distance, knowing and the unknowable, watcher and watched. So, not only does this work allude to the significance of movement when it comes to the object of surveillance, it also implicates the movement of the surveiller. In the kinaesthetics of the body required to engage in the aesthetics of the piece, Kemp implicates viewers in the politics and power inherent in the movement of bodies with access – bodies that inspect, scrutinize, and surveil. In engaging with the piece, the audience must determine the extent to which they have knowledge and access into the life of each individual through the information offered and how this personal data informs their understanding of that person. And as *Data Collection* makes clear, that access is never complete. The cards blacked out and marked as withheld are material markers of absented knowledge, a gesture to the incomplete picture of data itself. By emphasizing the gaps of knowledge, Kemp's work raises questions about how much we can actually know through viewing. As a proxy for the larger surveillant policing apparatus, the personal identification cards represent, simultaneously, access to knowledge and the gaps in this knowledge, along with the gaps in the knowledge-gathering abilities of surveillance technologies more generally.

The artworks in this section demonstrate the range and applications of different surveillant policing devices, from electromagnetic scanners, to airport X-ray machines, to identification cards. The widespread development, implementation, and acceptance of these technologies as information-gathering tools rests on the assumption that they present an objective and scientific basis for drawing presumptions of guilt, or soon-to-be guilt. While these artworks foreground the technologies themselves, I want to emphasize that they do so as a means to represent the less visible, more diffuse systems of authority, control, and oppression that dictate their use within the Canadian context. As these artworks illuminate, the knowledge produced through accessing and interpreting private information using these technologies is not neutral: it is defined in terms of the settler colonial Canadian project and the ways in which surveillant policing strategies uphold its logic. And as part of examining what these technologies reveal, these works deny the omniscience of the colonial state's vision and give substance to the spaces where surveillant policing cannot see: in Spriggs's *The Visible Spectrum*, all aspects of the image are never illuminated at the same time; in McQuillen's *The X-Ray Series*, the introduction of paper weapons skews the technology's capacity to read and assess threat; and Kemp's *Data Collection* allows volunteers to withhold any of their cards from public scrutiny. Each creative project features the limits to surveillant policing's knowledge as they think through the ways these techno-systems demand increasing transparency from the public while often obscuring the processes of perception and interpretation that take place behind the scenes.

This chapter has focused on creative works that generate a dynamic understanding of how surveillance is mobilized as a policing strategy to uphold the settler colonial project of Canada. These artworks critically and creatively address various systems of surveillant policing within the Canadian context to engage audiences in asking questions that generate new understandings of their applications. The eight works included in this chapter utilize the social and cultural familiarity of surveillant policing forms – whether classic detective equipment, bureaucratic documentation and identification, or sophisticated technologies – to spotlight, critique, and challenge them. While each individual work addresses specific meanings around surveillance logic as it relates to policing in the contemporary Canadian context, they come together to reveal how different modes of surveillant policing create visibilities that govern and assess bodies, people, and communities.

I developed this chapter's three themes – profile, access, and limits – to bring otherwise distinct works of art into conversation with one another, bringing to life specific themes and threads under the larger topic of surveillant policing. Although divided, the sections themselves are also meant to create knowledges and meanings between them. A recurring theme throughout the three sections is a challenge to the presumed objectivity and neutrality of surveillant policing technologies. Although this discussion is emphasized in the section on access, it relates to a conversation that starts in the earlier section on profile with regard to the artworks that challenge the presumed guilt of those caught in the surveillance lens; and this theme continues in the final section, with the creative works that illuminate the gaps in and limits to the knowledges gathered by surveillant policing technologies. Another thematic thread central to all the artworks is the use of embodiment as a strategy to resist the alienating and dehumanizing glare of the surveillant policing apparatus. The artworks bring to the forefront the humanity of those objectified and constrained by this apparatus, intervening to illuminate the gaps in knowledge in and the incompleteness of the prevailing narrative.

In all of these works, the artists position surveillant policing not only as a matter of institutional control enforced by recognizable police or military agencies, but also as larger and often more complicated and silent management systems that produce criminality and inclusion on terms determined by the settler colonial project, and its intersecting systems of oppression: racism, cisheteropatriarchy, and white supremacy. In this way, these creative works make present what criminologists Amanda Glasbeek, Mariful Alam, and Katrin Roots label as "the politics of not-seeing."[66] In their discussion of body cameras on police officers in Canada, they highlight that, while videos of policing may capture the actions of the police, they fail to reveal the underlying frameworks that govern how choices are made, why actions are taken, or how events are interpreted. In other words, while visualizations of surveillant policing can illustrate events on a basic, narrative level, they ignore the systems and structures that guide understandings of events. Therefore, the logics and belief systems upon which surveillant policing depend – Newell's "information politics of policing" – are often invisible or unseen.

The artworks in this chapter work to make visible how surveillant policing functions to monitor behaviour and to reaffirm the state's authority over space and identity. Circulated within the period of the imperial project of the War on Terror, these projects highlight a historical moment that saw intensified government assertions and corporate claims to the right to access and an accompanying denial of privacy to citizens/inhabitants as part of this security campaign. Yet, the rationale of the War on Terror is inseparable from the longer state-making project of Canada. The War on Terror upholds this project by extending colonial logics of control, security, belonging, and territorial domination under the guise of national and global security. In very different ways, the artworks here bring forward the unseen information politics of surveillant policing in their relation to buttressing the Canadian settler colonial. The creative projects reveal both the strategies and technologies of surveillant policing and also the logic and belief systems that dictate public understanding of their applications. They demonstrate that contemporary surveillance systems and policing strategies are not only shaped by, but also actively reproduce, the enduring structures of settler colonialism, and how they function as tools for asserting and sustaining state power.

Landing 3

For a white settler state such as Canada, land is power – physically, culturally, artistically.

Carmen Robertson, "Notes to a Nation"

This chapter examines artworks that explore contemporary discourses of surveillance, security, and public safety and their connections to histories of settler colonialism. Specifically, the eight creative projects in this chapter focus on the relationship between settler colonial narratives and surveillance logic through how these frameworks visualize and claim the land in Canada. Studies of surveillance often focus on the management of people – understandably so, given that people and their movements are obvious objects of interest when it comes to technologies and policies of surveillance. Certainly, the history of surveillance is anchored in the desire to control, monitor, police, and contain specific populations for a variety of reasons.[1] And, of course, studies of surveillance often focus on the places where people collect – particularly urban environments – reflecting the way surveillance viewing is tethered to monitoring bodies through cityscapes as microcosms of national inclusions and exclusions. I contend, however, that the history of surveillance is not one that exclusively involves people; importantly, it also centres land, resources, and extractions. Thus, there is space to look at surveillance through angles of address beyond people, which allows for consideration of the expansive reach of surveillance structures and their connections to settler colonial histories of broader terrains. In particular, Kevin Haggerty and Daniel Trottier point to nature as an understudied site of surveillance's reach: "We gesture toward a yet-to-be-written history of how nature has been configured by surveillance, and how nature is now positioned as a target, agent, resource, and model for surveillance practices."[2] "Nature" as a concept is fraught, with much critical scholarship thinking through its social, political, and historical construction as the absence of human civilization. Against persistent ideas of nature as "an island in the polluted sea of urban-industrial modernity,"[3] I find Haggerty and Trottier's gesture toward nature as an object of surveillance logic productive in thinking about surveillance beyond people.

In settler colonial societies like Canada, surveillance and colonialism are both tied to the idea of possession, which, in turn, is tied to whiteness. "Whiteness" here is not simply a racial category but, rather, is a dominant form of being in and seeing the world that is mutually constitutive with ideals of resource extraction, settler colonialism, and cisheteropatriarchy. This relationship normalizes gender and patriarchal norms that uphold white masculinity, heterosexuality, and the nuclear family unit as central to the structures of colonial capitalism.[4] Indigenous feminist scholar Aileen Moreton-Robinson uses the term "white possessive logics" to describe the rationalization of colonial desire that is "operationalized within discourses to circulate sets of meaning about ownership of the nation, as part of common sense knowledge, decision making, and socially produced conventions."[5] So, while settler colonialism is "motivated by access and control over territory," territorial expansion is "accomplished through the process of constructing white supremacy in relation to Indigenous inferiority," as education and social work scholars Amanda Gebhard, Sheelagh McLean, and Verna St. Denis argue.[6]

Surveillance visions of the land have been and continue to be integral to the possessive logics of settler colonialism and white supremacy. Studying the surveillance of Blackness, Simone Browne writes that surveillance "upholds negating strategies that first accompanied European colonial expansion and transatlantic slavery that sought to structure relations and institutions that privilege whiteness."[7] Systems of white supremacy, overlapping with other structures of oppression, are foundational to the settler colonial project, which is upheld by the logics and operations of surveillance. To think through the ways in which surveillance participates in shaping understandings of Canada as a settler colonial state, we must attend to forms of surveillance looking in relation to claiming land and territory. For settler colonial studies scholar Patrick Wolfe, "territoriality is settler colonialism's specific, irreducible element."[8] Therefore, to study the relationship between surveillance and settler colonialism in Canada necessarily requires a consideration of their mutual aim to capture, claim, monitor, and police the land.

Settler colonial claims to the land in Canada involve not only material possession but also visual and discursive possession. Art and architecture historian Carmen Robertson observes that "exploiting resources, tilling land, putting up fences, and parcelling out acres" leave visual markers on the land itself and produce a vocabulary of measurement and ownership, reflecting a "systematic process of claiming possession that contributed to the civilizing of the space that is the nation."[9] Such visual representations of land and white settlement are interdependent with the developments and expansions associated with settler colonialism, which itself relies on Indigenous dispossession. For Indigenous studies scholar and political scientist Glen Coulthard, this "colonial relation of dispossession" needs to be recognized as a "co-foundational feature of our understanding of and critical engagement with capitalism."[10] The visualization of these processes of dispossession contributes to white settler confidence in and authority over these colonial systems of capitalist extraction and conquest.

The visualization of land within Canadian national histories has a long tradition, one that is inextricable from the formation of the nation itself. Since the early nineteenth century in Canada, representations of land as available for colonial conquest and settlement became an established visual vocabulary that surveyed and surveilled territory. Art historian Marylin McKay categorizes this art tradition as a "nomadic concept of territory," whereby any land just beyond settled geographies is considered to be uninhabited and available for conquest. Such a concept supports "postcolonial societies in their search for material progress and nationalistic grandeur by encouraging the settlement of the wilderness and the extraction of its rich natural resources. Art that is based on the nomadic concept of territory naturally focuses on wilderness land and the active process of conquest and penetration."[11] The visualization of territory in the image of white settlerhood took the form of landscape depictions and was central to the processes of settler conquest and domination of the land. For instance, representations of early life in the European North American colonies reveal expansive uninhabited landscapes and sublime wilderness, often rendering invisible or symbolically small Indigenous life and settlements. Well-known artists such as Paul Kane (1810–1871) and Cornelius Krieghoff (1815–1872) painted expansive panoramas of supposedly "unsettled" land, images that marked and claimed the land as available for European occupation. Celebrated Canadian artists such as Tom Thomson and the Group of Seven famously centred white settler imaginings of uninhabited landscapes available for national expansion and colonial projects within their paintings. Canadian Pacific Railway posters, as well as government advertisements encouraging western and northern European immigration to Canada, foregrounded similar visual narratives of the land as uninhabited and available for settlement and extraction.[12] Representations of the land as almost or fully uninhabited have been central to narratives of Canadian-ness, and visualizations such as these played an active and specific role in what Indigenous studies scholar Andrea Bear Nicholas describes as "the colonial project of clearing the wilderness, settling the land, and establishing a new civil authority."[13] For Bear Nicholas, visual depictions of white settler life portrayed a certain image of colonial conquest that aided the material dispossession of Indigenous peoples and the conquest of the land. In this way, visual representations and material forms work together in the colonial project.

The creative projects in this chapter reveal the contemporary relationship between visual and material claims to the land and settler colonial histories, examined through the lens of surveillance viewing. They critically engage with and often draw from long-standing national art traditions that depict Canadian landscape as central to settler colonial dreams of territorial expansion and nation building, making explicit the connections between these colonial representations and the territories shaped by contemporary surveillance structures. The artworks in this chapter, however, question the very ideas of territory and landscape and what they reveal, not of the land itself, but of settler colonial expectations of the land, as they visualize how surveillance structures operate as an apparatus for territorial dominance. And while these works come together within this larger frame to denormalize the logic of settlerhood's claims to the land, each approaches its subject matter in different ways: some simply unveil the reciprocities between surveillance and colonialism; some intervene or make strange; some criticize outrightly; and some make more subtle interventions. All of the artworks in this chapter engage with representations of land and, I argue, expose the operations of surveillance as a vehicle for naturalizing settler colonial life on the land.

To look at the history of the people surveilled and surveilling is also to connect this history to the ways in which life and land are comanaged through systems of "white possessiveness" that define who has access and ownership to land and how such spaces are controlled. Produced during the period between 2001 and 2021, the works in this chapter came out of the period defined by the global War on Terror and the mainstreaming of surveillance as part of the everyday experience of increased security. These artworks address this moment of surveillance logic in its relationship to land, territory, and ongoing colonial narratives. In my analysis of these works, I ask: How does the settler colonial project take up surveillance to envision the land? What kinds of surveillance images have become widely representative as part of an accepted lexicon of colonial ownership? How do artists leverage the familiarity of these types of visualizations to challenge and rethink the relationship between surveillance, land, and settler colonialism in Canada? The works in this chapter envision the land with a surveillance lens, using the logic and technologies of surveillance looking to represent it. Read through a critically decolonial lens, these creative projects reveal the ways in which public spaces in Canada are marked, ordered, and territorialized. I examine these works to demonstrate the foundational connection between the logic of surveillance and settler colonialism – that is, how surveillance viewing is settler colonial viewing – and the ways in which this connection is central to historical and ongoing projects of colonial settlement in the Canadian territory.

This chapter is organized around three main curatorial themes that consider surveillance operations on the land: mapping, private property, and landscape.

In "Mapping," the artworks address how mapping and charting visualize knowledge and claim territory; in "Private Property," they ask how social relations are reproduced via property ownership and protected by practices of surveillance; and in "Landscape," they expose surveillance forms of looking that attempt to confirm colonial mastery of land. Each of the works in this chapter proposes a different angle of vision on surveillance logic and its structures and challenges colonial imaginaries that produce and reproduce settler colonial life on, and extraction of, the land. In bringing forward the logic of surveillance in relation to projects of settler colonialism, these artworks connect territories of contemporary surveillance structures and representations of colonial expansion and raise questions about for whom the land is marked and protected. Taken together, the works highlight the reciprocities between settler colonialism and surveillance and seek to dismantle and denormalize the logics of white settlerhood's claims to the land.

Mapping

Surveillance in its relation to charting and plotting the land is a vehicle of ownership, authority, and control within settler colonial systems. In early Canada, surveillance was used in the service of charting, and, as political scientist Ahmad Sa'di notes, "colonial administrations used cartography as a tool of governance."[14] This strategy inextricably links colonial visions that surveil the land to capitalist extraction and imperial imaginaries. The surveillance lens on land and space renders place as property – public and private – and creates parameters around who and what can inhabit the land and to what ends. The artworks in this section consider how settler colonialism demarcates and claims national spaces through mapping as a way to envision its own image. As cartography historian and geographer Matthew Edney articulates, "imperialism and mapmaking intersect in the most basic manner. Both are fundamentally concerned with territory and knowledge."[15] For colonial settlers, maps function as a tool of colonial-capitalist knowing, a way to affirm and visualize settler perspectives of and claims to the land. In this way, land is transformed into territory.

Yet mapping is not only about the spatial outlines of a geographic area; it is also about the imagined connections to space that demarcate the land in terms of emotions, memories, nostalgia, and particular historical narratives. These aspects, too, often enable colonial attachments to and control of the land through a constructed context that orients the uses of the land and the lives on the land around white settlers. For political geographers Mia Bennett, Wilfrid Greaves, Rudolf Riedlsperger, and Alberic Botella, "maps are crucial instruments for defining a region and affecting popular perceptions and understandings of its geographic, social, and political shape."[16] Maps provide a familiar and comfortable touchstone for shared understandings of space – a path to claim collective ownership and knowledge over the organization of the land. According to Indigenous studies scholar Mishuana Goeman, mapping becomes a tool that "structure[s] settler colonial geographies" and "put[s] into place settler understandings of geography."[17]

Like art, maps represent and visualize the land from a particular perspective. Maps are a prominent way in which land and territory are re-envisioned as settler property and resource, not only for what they render visible but also for what they leave invisible or unmarked. The three artworks in this section probe settler colonial perceptions of the land as owned, propertied, monitored, and collected, yet they also provide different conceptions of how the land is imagined and represented, or not represented, in colonial mapping projects. The creative projects here approach the broad idea of surveillance as a tool for settler mapping and geography in the Canadian context in different ways, providing multiple perspectives and entry points to think through human relations to the land via surveillance practices. They connect in their shared interest in perceiving surveillance operations as a cartographic device as well as in their decolonizing approach to moving beyond settler imaginaries of the land and territory.

Hiba Ali's installation *Satellites* (2016, fig. 3.1) approaches surveillance mapping's "systems of vision by investigating landscapes known as frontiers, both 'outerspace' and 'western,' digital and physical, metaphoric and literal."[18] Organized by the South

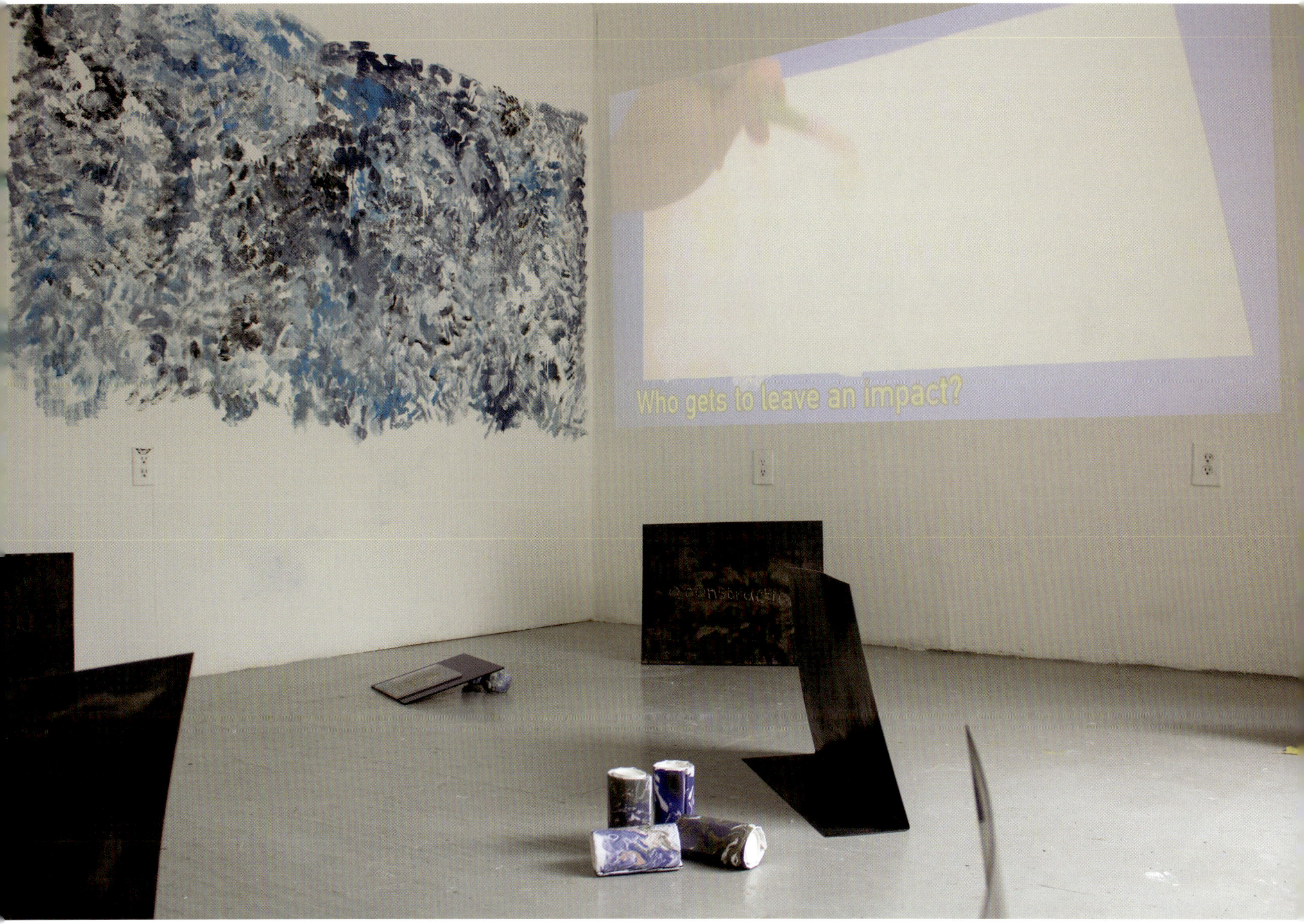

3.1 • Hiba Ali, *Satellites*, 2016.

Asian Visual Arts Centre and exhibited at the Whippersnapper Gallery in Toronto, from 6 to 27 June 2017, *Satellites* explores the relationship between geospatial images made by orbiting satellites, and Eurocentric colonial conquest and perceptions of land.[19] The work is multimedia in design, composed of sculptural installations of interpretations of the globe, wall-mounted paintings that reimagine visions of the earth and the land, and video projections of computer-generated interpretations of Earth's surface, along with dissonant music, spoken words, and written text. The audio component of the video projections contextualize the intent and aim of Ali's overall project through questions and statements that point to the way images of Earth inform our imagined place on it: "How do we imagine the world?"; "The imaging of satellites and infrastructure enable globalization and … disseminate the models of Western Eurocentrism"; "These systems visually and sometimes literally tell us how to imagine ourselves within the world around us"; "Can we separate imperialism from our minds?"; and "Is space something to be conquered?" These pronouncements interrogate the settler colonial vision of the land for what it creates and projects, and also for how it is circulated as the dominant understanding.

Ali further reinforces the importance of asking how our perceptions of the world are formed when they pull back the curtain on how they went about creating their images. In some sections of the video projections, Ali records the process by which they created some of the paintings that appear in the installation, which involved stamping the canvas quickly and firmly with a sculptural interpretation of a globe. By making impressions on the canvas in the form of Earth's image, Ali brings together the making-process and content of the work, imagining the manner in which meanings from images impress on our minds. In *Satellites*, Ali connects the technologies of surveillance geospatial devices circulating the globe in outer space to Eurocentric imperial visions of Earth as a resource for accumulation, extraction, and capitalist growth; they then challenge these dominant perceptions by questioning where they come from and who gets to make them.

Ali treats outer space travel and mapping not as forms of unbiased surveillance communications and information gathering but as extensions of the settler colonial process that shapes wider understandings of the land in terms of ownership, inhabitation, and control. In so doing, this art project reveals the underlying logic driving surveillance visions of Earth and its lands as constructed in the image of Eurocentric imperialism – that is, that outer space exploration is an extension of colonial conquest. This colonial logic dictates much about how Earth and its resources are allocated and used. As information studies scholar Réka Gál articulates in relation to space exploration programs lead by rampant industrialists such as Elon Musk, "space exploration is explicitly settler colonial. It projects the same logic of *terra nullius* into outer space that was used as a justification for the appropriation and colonization of the North American lands that were inhabited by various Indigenous nations, while also reproducing existing colonial relations on Earth through the expansion of space colonization infrastructure."[20] *Satellites* takes satellite mapping that seeks to represent and "make known" the intricacies of Earth's surface and reveals the logic that lies beneath the surface.

Ali's reinterpretations and altered angled visions of Earth are not merely aesthetic reimaginings; instead, they are an investigation into how and why our knowledge of this planet and its mapped contours was produced in the first place. Ali exposes planetary mapping as a tool of conquest, domination, imperialism – representationally and also psychically – as they reframe satellites from mundane and neutral tools of observation and record making to nefariously surveillant. They show how planetary mapping carries on the imperial spirit of earlier forms of colonial expansion and serves larger interests of capitalist extraction. Mapped images of the land are produced through the lens of this logic, even while they are circulated as the dominant and objective mode of understanding Earth. Ali seeks to dismantle this vision by "unearthing" and spotlighting this perception, and asking the audience to think through their own understandings and complicities in its construction.

While Ali's *Satellites* positions mapping from a distance via satellites, Jon Rafman's *Nine Eyes of Google Street View* (2008–ongoing) examines surveillance mapping technologies from street-view level. Photographs taken from the project have been published in a book under the same title and shown in several exhibitions, including *Cyber-Surveillance in Everyday Life*, a group exhibition at the InterAccess Electronic Media Arts Centre in Toronto in 2011. *Nine Eyes* is a photographic series that extracts a curated selection of still images from Rafman's extensive searches of Google's Street View. With this work, he translates the interactive interface of Google Street View into still photographs, like *Auburn Bay Blvd SE, Calgary, Alberta, Canada* (fig. 3.2), with recognizable aesthetic associations. Rafman uses a variety of criteria for which images he selected as part of this project, including unedited pictures of urban and rural life; images that recall the aesthetics of historical photographs of family photos; and scenes that mimic compositions of well-known art photographs.[21] The artworks in this series function to defamiliarize the collection of mapping images produced by Google Street View by recontextualizing them and forcing viewers to rethink their understanding of how and why Google generated the images and what meanings they produce.

While the images themselves produce particular meanings, Rafman's extraction of these images from Google Street View lends a specific context to the production and circulation of *Nine Eyes*. Launched in 2007, Street View is a technology intended to provide interactive panoramas of streets around the world and is part of the worldwide mapping initiative of the multinational tech company Google. Originally involving nine cameras mounted on the top of a car moving slowly through car-accessible streets – a process highlighted in the title of the artwork – Street View now spans seven continents, using 360° cameras mounted on a range of mobile vehicles to offer, "from the privacy of one's networked screens," the ability to "access a seamless, panoramic map of digitally stitched photographs taken from street-level all over the world."[22] Communication scholars Chris Ingraham and Allison Rowland assert that this program is "one of the most comprehensive surveillance mechanisms in human history."[23] Google presents Street View as a neutral technology, but, as American studies scholar Jason Farman writes, "the delineation of borders and the naming of territories have such historical intimacy to the control that empires wield [that] Google's sustained interest in digital maps have made them a key node of command over the 'information empire.'"[24] This "information empire" is anchored in

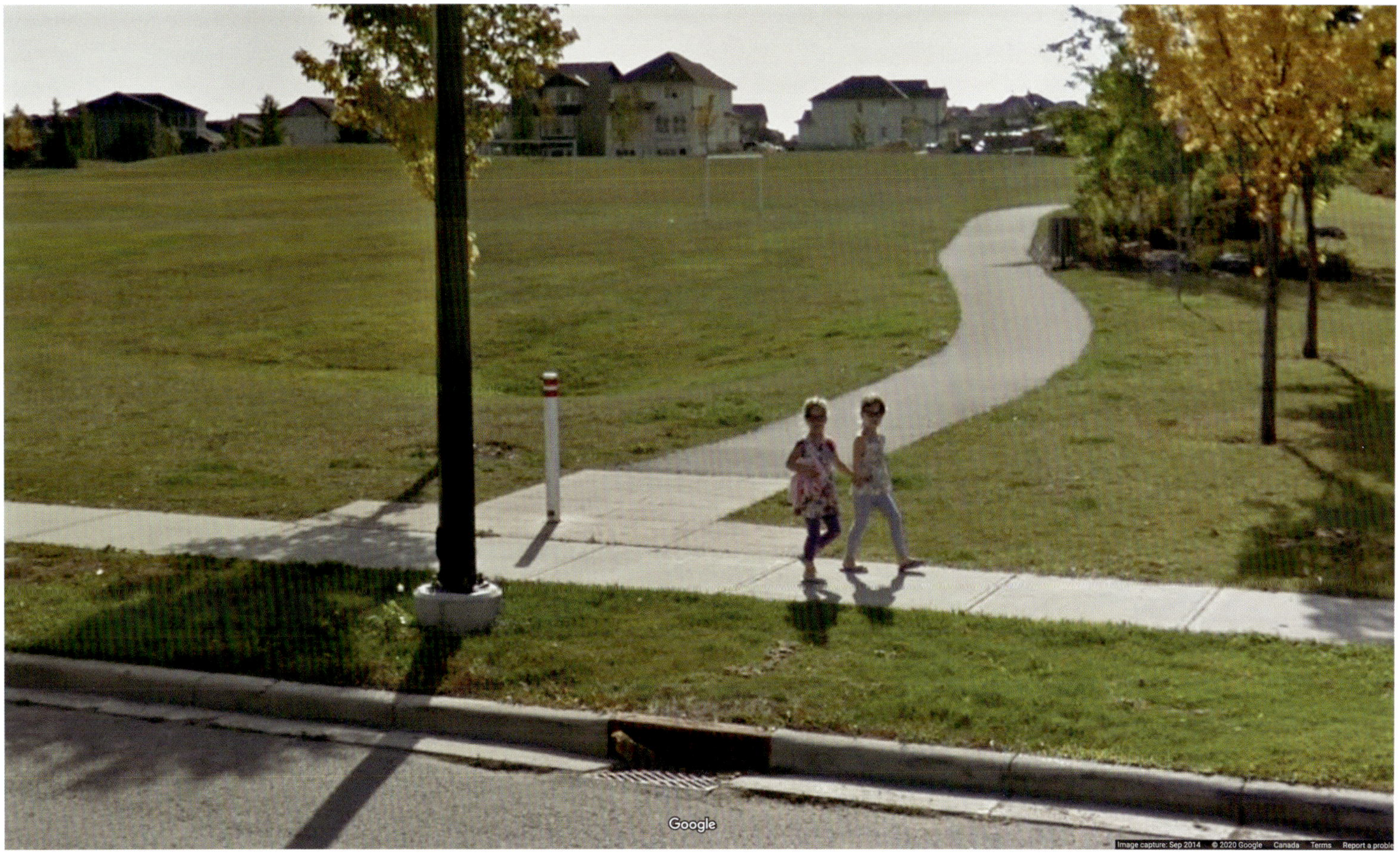

3.2 • Jon Rafman, *Auburn Bay Blvd SE, Calgary, Alberta, Canada*, 2020, from the series *Nine Eyes of Google Street View*, 2008–ongoing.

extraction, control, and authority – a sense of the power to be the one who gathers, monitors, records, and processes information related to land and life on it. Google's own mission "to organize the world's information and make it universally accessible and useful" attests to this.[25]

In Rafman's own words, the extraction of images from this enormous online archive challenges "Google's imperial claims" and "the company's right to be the only one framing our cognitions and perceptions."[26] This artwork alerts us to Google's mass accumulation of data about both the land and the use of that land, and interrogates how this collecting, curating, and interpreting are forms of power and control. It connects this contemporary digital mapping project to longer histories of claiming and visualizing the land as part of empire building and foregrounds how the data and information produced through these images enable forms of capitalist extraction and colonial expansion. By freezing and fixing the expansive, interactive consumption of land and life offered by Street View, the artwork interrupts the viewer's habitual scanning and movement. It slows down the user's ability to absorb the landscape as a seamless whole, redirecting attention toward the specificities of what each image captures.

Importantly, *Nine Eyes* helps illuminate the ways in which a project like Street View heralds a new cycle of dispossession provided by digital colonialism, one where the traditions of colonial mapping of the land through Cartesian coordinates is tethered to vicarious visual experiences of being on that land. As film and communications scholar Ross Gibson writes, "Google Street View ... links 'on-the-ground' landscape renditions to traditional coordinate-charting via GIS surveys and networked user tracking."[27] In so doing, it conflates viewing the land at a distance with experiencing and being on that land, and it renders invisible the mapping of colonial ownership. Scholar of technology and psychology Shoshana Zuboff points out that it is Google's prerogative to "empty every place of the subjective meanings that unite the human beings who gather there."[28] Such a prerogative is tethered to logics of capital possessiveness and the spatialities of the colonial geographic imaginary. What Street View purports to offer to viewers/users is the capacity to become a connoisseur of the life and land of a place without the context of the life lived there and being-in-relation to the land.

The interactive images made still by *Nine Eyes* recontextualize this experience as art and demonstrate the distance between viewing and being in place on the land. Certainly, *Nine Eyes* highlights "the ethical and political dynamics of critical cartography, characterised by a tension between *reinscribing* and *undoing* colonial power."[29] By selecting images that recall established photographic aesthetics, Rafman creates a tension for audiences: the images feel formally familiar, even as they unsettle the viewer's relationship to what is depicted and the conditions under which the images were produced. The title of the artwork, *Nine Eyes of Google Street View*, emphasizes the mechanisms through which these images are captured – the nine eyes mounted atop a vehicle – and forefronts the surveillance operations that yield these mapped landscapes. It encourages viewers to ask what type of knowledge is gained through the distanced viewing of land and the life on it. What is rendered absent in these images? What gaps in knowledge are produced through distanced viewing? What logic underlies the presumption of knowing and experiencing a place vicariously and mediated through Google? In this way, *Nine Eyes* operates as a counter-cartography project, one that works to illuminate the colonial and surveillance logic of Google and its Street View technology. To me, this art project works in conversation with other counter-cartography projects, which, as curator and activist André Mesquita asserts, seek to resist and rethink imperial and colonial mapping in order to make domination visible.[30] *Nine Eyes of Google Street View* brings forward a critical challenge to how technologies and perspectives of surveillance intersect with settler colonial imaginaries of mapping the land as a resource for understanding, ownership, development, and management.

Art projects that interrogate surveillance in relation to mapping the land do not always focus on the geographic. Some open up possibilities to capture the emotional or psychic aspects of landscape through personal experiences and histories of the land. In 2013, Vancouver's grunt gallery invited Dutch artist Desiree Palmen to recreate a previous art project she had undertaken in Turkey called *Saklambaç* (Turkish for "hide and seek"). Reframed to fit the local context, Palmen's *Mamook Ipsoot* (Chinook for "to hide or make hidden") involves seven local Indigenous artists, ages eleven to fifteen, who volunteered to collaborate with Palmen: Danniel Acosta, Laticia Baptiste, Jaden Jugovic, Carson Pechawis, Sage Markham, Dene Couture, and Raven Kahnapace-Bird. In this project, these youth chose urban locales in which to camouflage themselves. Palmen assisted them in selecting a site, developing a camouflage costume, and photo-documenting the results. The project coordinator, Jolene Andrews, describes the photographs as being from "a birds eye view, as if they were taken from the cameras that safeguarded the city"[31] – in other words, the photographs of the young artists in situ are intended to mimic the surveillance gaze of security cameras. The logic of surveillance in its relationship to settler colonial imaginaries of the land underscores this project.

The goal of this work, like multiple past projects led by Palmen, was to intervene in the spaces of everyday life.[32] *Mamook Ipsoot* provides new ways of understanding place and land, with the lenses of young Indigenous artists and their emotional relationships to place shaping the perspectives. The process of creating these images involved each artist selecting a place that had special meaning to them and photographing that site. Then, donning a white shirt, pants, and hat, the artists placed themselves within their chosen sites and photographed the site a second time. Once these reference photographs were taken, each artist worked on painting their white outfit to blend with their chosen site. The final part of the artistic process was to return to their respective sites with their painted outfits and to position themselves in the same location as their reference photos.[33] The resulting images reveal the artists and their camouflaged bodies blending into place. The camouflage becomes symbolic, emphasizing the aspects of life that cannot be monitored or controlled through the surveillance lens, the gaps and absences of interior life, and a counter-narrative to the foundation of settler colonial surveillance logic as a form of looking, knowing, and controlling the land.

The youth were encouraged to select spaces that had meaning to them. One of the artists, Jaden Jugovic, selected the Scotia Bank Theatre in downtown Vancouver because he likes to watch movies there, while another, Laticia Baptiste, selected BC Place because its location near the ocean allows her to look over the water.[34] In this way, the resulting images function to punctuate the affective qualities of the urban environment in their work. In Carson Pechawis's photograph, entitled *Invisible in Mother Nature* (fig. 3.3), a circle of large rounded stones surrounds a small shallow pond. Tall ornamental grasses grow around the stone border, creating a boundary around the site that separates it from the trees and houses visible in the background. To the left of the image, the artist herself is barely visible, perched on one of the stones and facing the water. Pechawis's shiny brown hair is what first gives away her presence. On closer inspection, the viewer can spot the painted suit: the top coloured to disappear into the grass and the bottom blending in with the stones. It is a peaceful and contemplative scene, seemingly disconnected from the expectations of what one might see in an urban centre like Vancouver. Pechawis selected this tranquil setting in the middle of downtown Vancouver because it allows her to feel connected to the calm and quiet.[35]

3.3 • Carson Pechawis, *Invisible in Mother Nature*, 2013. From Desiree Palmen, *Mamook Ipsoot (To Hide or Make Hidden)*, 2013.

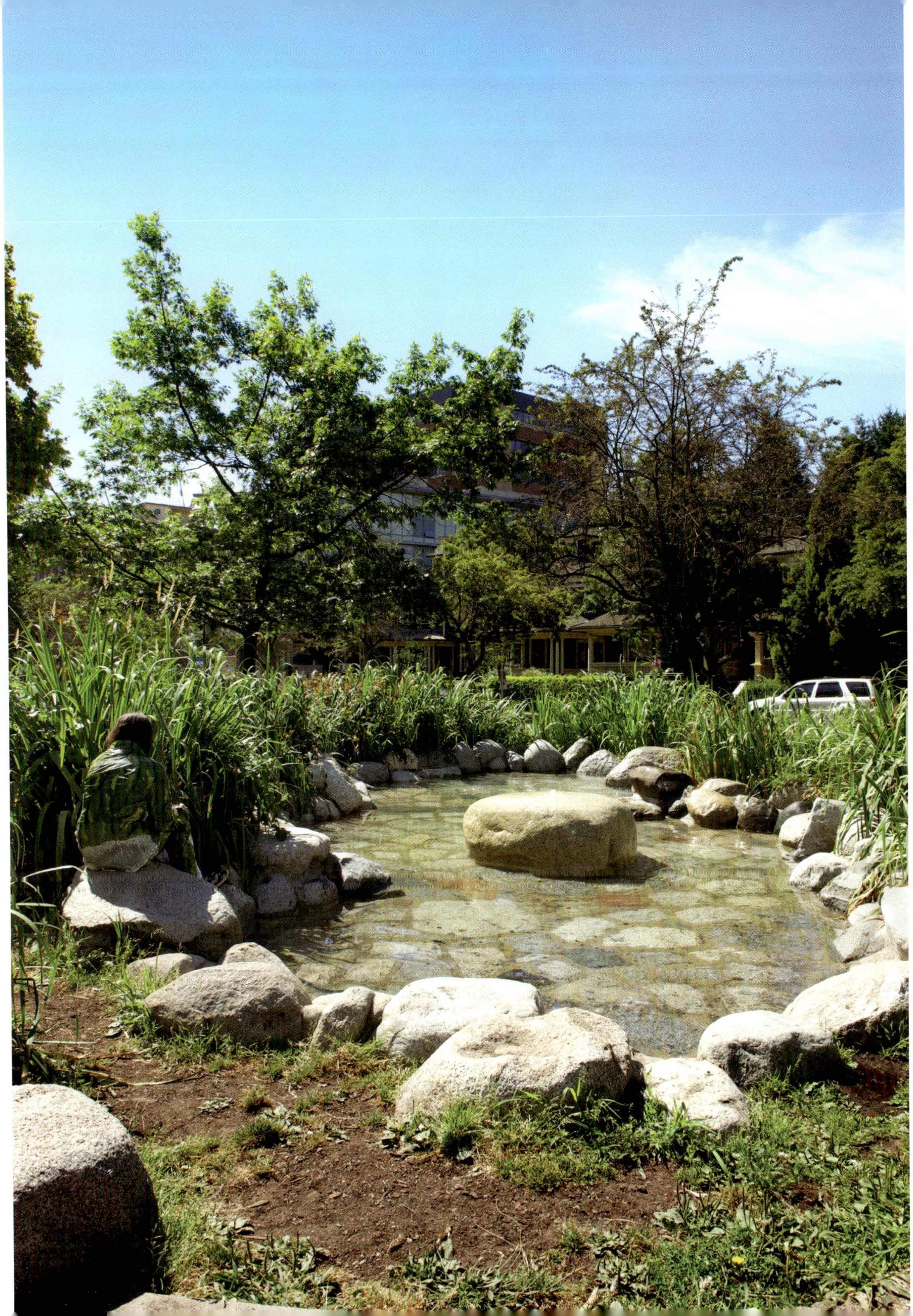

The form of mapping in *Mamook Ipsoot* prioritizes personal, reflective, and emotional ways of knowing and relating to place, in contrast to the scientific, authoritative, or objective interpretations associated with geographic mapping. Mapping in this sense becomes quiet and reflective, requiring personal narration to make meaning. Such a strategy reveals the shortcomings and absences of surveys and surveils of the land that seek to subsume the landscape and its multiple histories and meanings into a simplified and widely discernible understanding of space. Taking place in Vancouver, on the unceded territories of the xʷməθkʷəy̓əm (Musqueam), Skwxwú7mesh (Squamish), and səlilwətaɬ/selilwitulh (Tsleil-Waututh) Nations, *Mamook Ipsoot* provides an alternate mapping of the land, one that is anchored in the young Indigenous artists' own experiences and understandings of place. This project's approach stresses the gaps and non-understandings inherent in surveillance that is used to try to track and contain the land and its uses. In doing so, it underscores how wide-scale monitoring fails to register the intricacies of place and the relational ties between people and land.

The emphasis on camouflage points to the hidden and unseen life within urban communities. Intended to fully incorporate the young artists into their chosen locale, this meticulous façade also works to question the logic of who is seen or unseen, in what ways, to what ends, and in what spaces. Certainly, the focus on Indigenous youth in this project lends a perspective and meaning to the work that is connected to ongoing histories of colonialism within the Canadian context that surveils, monitors, polices, contains, and controls Indigenous people, communities, and life in antagonistic and violent ways. The camouflage blends the bodies of the artists into particular locales that feel emotionally resonant to them and asserts their claims to and custody of place. It also sets a playful tone, because it helps the title of the artwork come to life as the childhood game of hide and seek. Yet, at the same time, the camouflage thwarts this playfulness, as it points to the historical and ongoing surveillance of Indigenous life on this land and the ways in which the state and its forces have sought to violently eradicate Indigenous existence. Camouflage in this sense becomes simultaneously a gesture of hiding – a way to avoid the surveillant colonial gaze – and one of asserting Indigenous life and presence in place and on the land.

Mamook Ipsoot speaks to Indigenous resilience and the ongoing and vibrant presence of Indigenous youth and communities. Although facilitated by a non-Indigenous artist, this project foregrounds the seven collaborating artists to narrate their relationship to the urban environment. Small, personal vignettes of their chosen places and their attachments to them thread through the different images, providing counternarratives to "the danger, both emotional and physical, of surveillance of visibly racialized Indigenous bodies, surveillance that occurs everywhere, from public streets to their own workplaces."[36] In this way, *Mamook Ipsoot* calls to Indigenous futurity on the land and rejects the dominant white settler colonial imaginary. Acosta, Baptiste, Jugovic, Pechawis, Markham, Couture, and Kahnapace-Bird, in collaboration with Palmen, chart their experiences of land, space, and community by selecting emotionally resonant locales and then blending themselves into these environments. The resulting images were exhibited on bus shelter posters and in various off-site locations around Vancouver before being installed in the exhibition space of grunt gallery. Their installation in public spaces helped further affirm the presence of these young artists and their works in the environments and communities around them, captured in the images. The images and their public exhibition leave traces of these seven Indigenous artists as fixtures within the landscape, present whether others can see them or not.

The three artworks in this section invite viewers, first, to question the ways in which surveillance images inform how they imagine their position or place of belonging on Earth or on land; second, to notice that the so-called neutral images of unimpassioned mapping photography in fact create cultural texts and social meanings, raising the question of who should have such power; and third, to resist surveillance's attempts to influence emotional ways of knowing and relating to place. Together, they reveal what is made invisible in the normalization of a surveillant colonial perspective. Ali's *Satellites*, Rafman's *Nine Eyes of Google Street View*, and Palmen's *Mamook Ipsoot* – with the works by Acosta, Baptiste, Jugovic, Pechawis,

Markham, Couture, and Kahnapace-Bird – present different visions of thinking through surveillance's encounter with land, geography, and mapping. These three creative projects present ways of revisualizing the settler colonial imaginary by challenging the dominance of mapping representations as authoritative, objective, and static, and they do so by navigating surveillance in its attachment to colonial-capitalist extractions of the land as data, as resource, and as a contained entity. To this end, these artworks each – and in dialogue with one another – bring to life what social scientist Ingrid Waldron asserts about the relevance of geographic space: that it is "a socially constructed and highly contested product and idea bound up in political, cultural, and economic meanings."[37] By creatively and critically imagining different landscapes as part of a decolonial project that rejects one way of seeing, understanding, and controlling the land, these three creative projects are united in revealing not only what is made visible, but also, significantly, what is made invisible, unseen, or marginalized through the surveillance lens of settler colonial mappings.

Private Property

The works in this section focus on the intersection between surveillance logic and settler colonial claims to the land as private property. These artworks focus on one of the ways settler colonialism demarcates and visualizes the land in its own image: they reveal how colonialism lays claim to the land by treating it as a resource to be exploited for settlerhood's own interests, and, in so doing, they materialize how, in Aileen Moreton-Robinson's words, "signs of white possession are embedded everywhere in the landscape."[38] By focusing on different iterations of private property and its management through surveillance structures, the two works in this section function as a critical rethinking of the reciprocal relationship between surveillance and colonialism. Together, they reveal how these systems produce and sustain settler colonialism's naturalized relationship to the land, and its capitalist, propertied claims upon it.

Colonial capitalism within settler colonial states is premised on the right of certain people to claim ownership of land. As American studies scholar Alyosha Goldstein highlights in relation to US settlerhood, "property relations and prevailing conceptions of ownership are key sites for the social reproduction of race, capitalism, and the particularities of U.S. settler colonialism."[39] This claim is true within the Canadian context as well. The right to access and possess territory is an important way that liberal nations like Canada define who is recognized as possessing democratic rights and citizenship. The accumulation of and attachment to territory is how "whiteness accumulates capital and social appreciation as white people are recognized within the law as property-owning subjects." For that reason, white people are "heavily invested in the nation being a white possession."[40] Surveillance, in this sense, becomes a policing tool that establishes and maintains white settler life on the land. The artworks here position claims to property ownership within this context of establishing subjecthood and citizenship, and ask how the surveillance of private property functions to monitor, police, and control the land and the people on it as part of the reproduction of settler colonial systems.

Stéfy McKnight's installation *hunting for prey* (2016, fig. 3.4) raises many questions about the relationship between humans and more-than-humans on the land and the role of surveillance in monitoring them. *hunting for prey* is part of a larger project, *Organic Surveillance*, that grew out of McKnight's master's thesis. Exhibited at Queen's University in Kingston, Ontario, in November of 2016, this creative project centres on McKnight's interest in what they identify as the "function creep" of hunting cameras toward a home-surveillance device. Hunting cameras were originally intended to record and track the movements of animals on the land, although they have been repurposed to record and track the movements of potentially unwanted people on the properties of landowners in rural northeastern Ontario, where McKnight is originally from.[41] For this series, McKnight interviewed local residents about their use of these cameras and gained permission to take photographs of the cameras mounted around these residents' properties. Additionally, McKnight bought their own infrared hunting camera and took photos on their own family's property

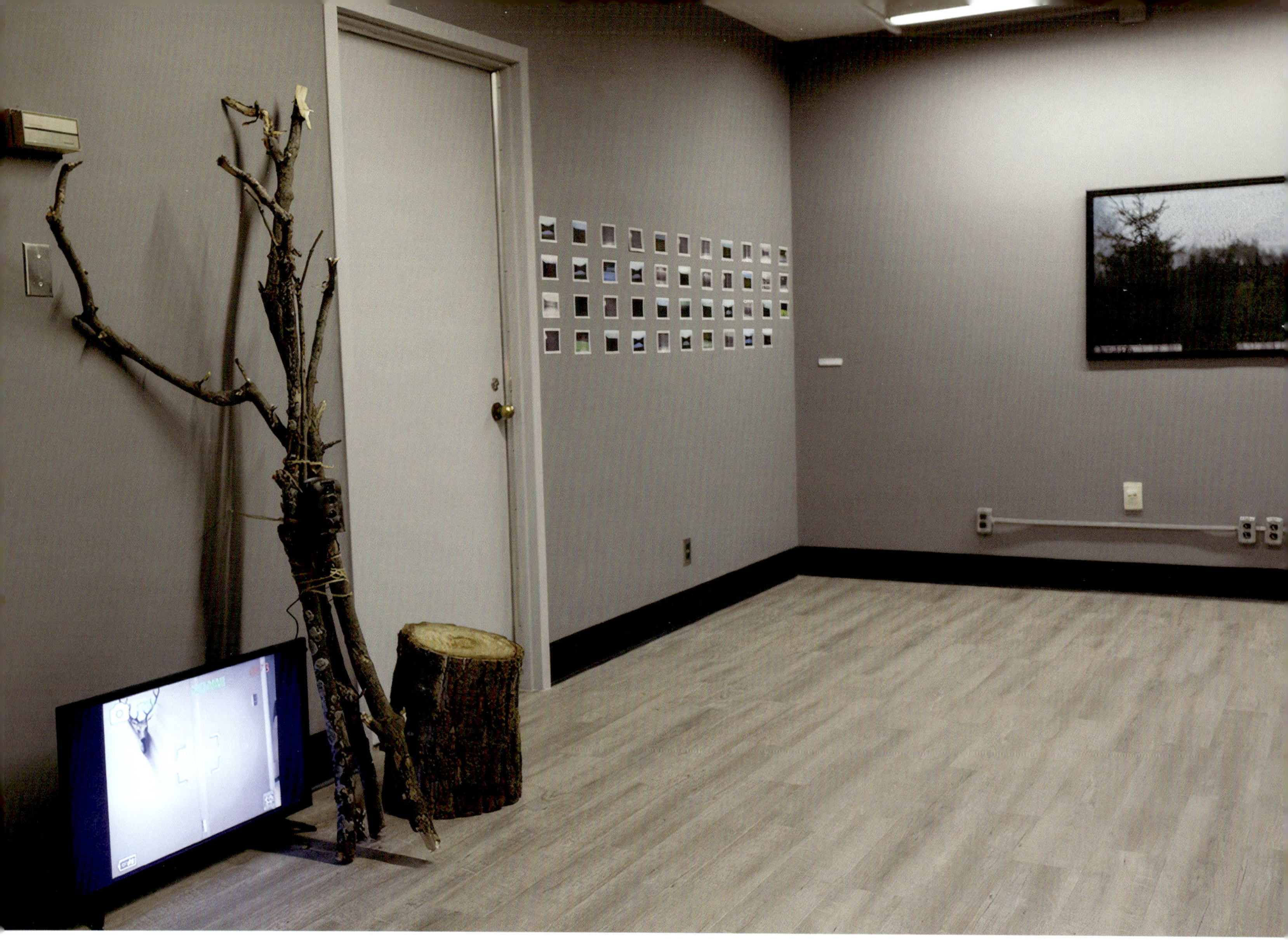

3.4 • Stéfy McKnight, *hunting for prey*, 2016.

at night over a period of time. The resulting photographic series installed in the exhibition space includes images produced by McKnight's own camera set-up – printed as Polaroid-like photographs and mounted in a grid on the gallery wall – as well as large-size images from the residents' hunting cameras in situ, which highlight the visuality and presence of these cameras in the rural Ontario environment.

The installation invites visitors to enter the space and engage with the photographs of the land, animals, and natural life captured by the cameras, although the gallery viewers are geographically removed from the objects of surveillance. From a distanced vantage point, it would be easy to see a rural folksy-ness in this project, a charming collection of photographs re-envisioning surveillance logic to provide insights into underseen aspects of natural life on the land. But another aspect of the installation forecloses any perceived distance, by implicating the body and movements of the visitors within the gallery space in the work itself. Mounted atop an installation of a tree stump and branches resting vertically up the gallery wall to mimic the tree landscape of northeastern Ontario forests is a camouflaged hunting camera. This camera is the same one McKnight used to produce the night-time images from their family's land. In the exhibition space, it records the movements of visitors in the gallery and projects these images onto a flat-screen television sitting on the floor at the base of the branches. Here, McKnight mimics processes of property surveillance in rural Canada, whereby cameras originally meant for tracking and hunting animals have become repurposed by landowners to surveil, monitor, and guard their land from unwanted and uninvited human interlopers.

McKnight's *hunting for prey* brings to light in clear and often parodic ways how the function creep that turns a hunting camera into a home-surveillance device to protect private property extends surveillance logic into new terrains. As McKnight has articulated, the increasing presence of hunting cameras used as surveillance devices reflects the encroachment of surveillance vision into previously under-surveilled areas.[42] It also reflects the broader availability of surveillance equipment to wider areas of society, serving to maintain settler colonial belief systems of the land as propertied and controlled by settlers. Importantly, *hunting for prey* shows that surveillance logics that secure claims to the land as owned and propertied are found not only in urban spaces but also in sparsely populated rural spaces. In this, McKnight's installation materializes what Haggerty and Trottier identify as a misconception that "nature" can be an idealized escape from surveillance measures: "In a world where individuals are repeatedly exposed to and by surveillance measures, one occasionally encounters a sentimental yearning to escape all this scrutiny. At the most extreme, this entails appeals to go 'off the grid'; to retreat into nature in order to extricate oneself from the institutional routines for which surveillance has become a precondition."[43] Certainly, McKnight's artwork illuminates how technologies of surveillance uphold and bolster colonial settlerhood's imaginaries and claims to the land as property, while reproducing the concept of "natural environments" or "nature" as constructed for the use of some to the exclusion of others, an ongoing hallmark of colonial expansion within the Canadian context.

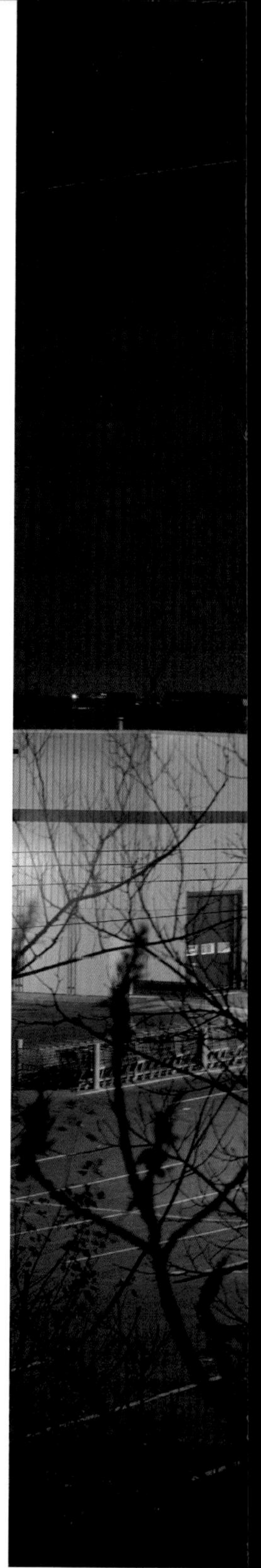

3.5 • Thomas Kneubühler, *Private Property #3*, 2006.

Thomas Kneubühler's series *Private Property* (2006) bears similarities to and key differences from McKnight's *hunting for prey* in its creative approach to ideas of space, access, property, and surveillance. Exhibited in the FOFA Gallery at Concordia University in Montreal in 2007, *Private Property* is a series of photographs of private corporate property and the personnel tasked with securing it. It comprises several exteriors capturing the spaces around corporate buildings, and a small number of portraits of the security guards inside the buildings, both set at night. The exteriors depict empty parking lots, the outside of lit buildings with drawn blinds, and shadowed cityscapes with skylines marked by dim office buildings wrapped with ribbons of interior lights running horizontally around them. These representations of the external structures of office life are made dramatic and uncertain through the interplay between the bright lights of the interior spaces and the darkness outside them where the photographer, and thus the viewer, is positioned. Kneubühler's framing of these landscapes often juxtaposes stark and austere office buildings and surrounding parking lots with hints of greenery: nature at the edge of capitalist workings. As in *Private Property #3* (fig. 3.5), many of the photographs take the viewpoint of looking at these corporate spaces through trees or from a downward perspective

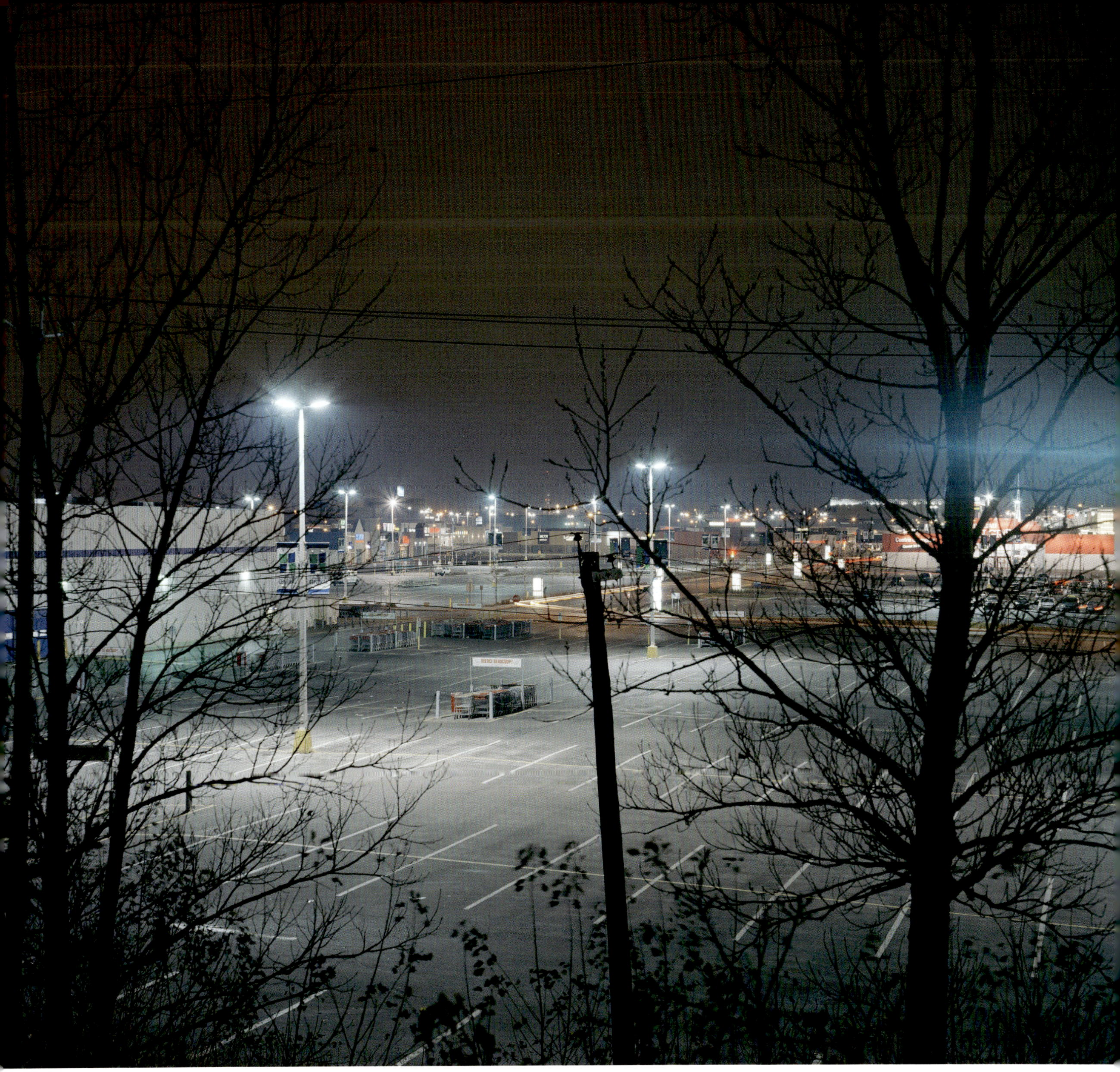

from a grassy, wooded hill so that the photographic framing of the images themselves mimics angles of surveillance viewing in their depiction of the spaces of commerce.

Several of the photographs invite the viewer inside the corporate space, but only conditionally, their access mediated by representations of the security apparatus that monitors and protects these structures from public viewing – namely, the security guards and their equipment. Photographs of security guards posed in the fluorescent glare of a hallway or under a bright street light in a parking area provide context from which to view the series as a whole. The handful of photographs of guards set against the array of impersonal exteriors underscores the surveillance operations that maintain and protect corporate life and workings from wider public access. One interior image depicts a wall of CCTV cameras, stressing the extensive technological operations and communication systems that complement the role of guards on patrol, and referencing the omnipresence of technologies that monitor movement and access. The portraits of the security personnel frame the guards as the sole human presence within empty, often lonely spaces, and emphasize the bodies of those who are often unseen behind the camera. In these ways, *Private Property* explores the relationship between the personal and impersonal in various systems of security and surveillance and examines who and what commands the authority to watch and control.

Like McKnight's piece, *Private Property* raises questions of surveillance logics tied to property, ownership, containment, and access. For Kneubühler, there is a firm boundary between insider and outsider knowledge and access: devices such as lighting, viewpoints, and the obstructing of interior views with the bodies of security guards establish firmly that these corporate spaces are not available to the public. Surveillance here operates to deny intellectual and physical access to the corporate life of these structures. In this denial, this artwork illuminates the wide swath of urban areas that are contained away from public viewing, with extensive security structures normalized as necessary to protect and maintain the public good of private industry. As professor of criminology and security studies Allison Wakefield writes, "the privatisation of urban space has raised concerns among many commentators, as a process that systematically sorts the privileged or compliant from the undesirable or disobedient."[44] The frame of these images creates a context that is very "outside looking in," preventing access into the inner workings of these corporate spaces beyond the posed portraits that centre the figures of the security apparatus that monitors and contains them.

As the title of this photographic series implies, the privatization of urban space holds privacy and claims to property in opposition to public life and to public knowledges. Property – and the wealth, knowledge, and power attached to it – is tethered to the workings of surveillance and security insofar as each upholds and maintains the other. With this work, Kneubühler materializes a consequence of what technology and privacy scholars Bennett Cyphers and Gennie Gebhart call "the one-way mirror" of corporate surveillance: companies and organizations establish infrastructure to monitor the public but are limited in how they themselves are regulated.[45] Many contemporary studies of surveillance focus on the digital aspects of corporate surveillance, while *Private Property* presents the analog version of the same surveillance logic that entitles corporations to claim space, demarcate it as property, and monitor and surveil anyone who crosses its boundaries.

In highlighting the insider and outsider access of these spaces, Kneubühler's piece asks questions of belonging more generally. I see this work as linking the microcosm of specific corporate property to macro issues around who gets to claim land and thus define the Canadian nation-state. Kneubühler illuminates the boundaries between inside and outside in *Private Property*, using light and darkness to literally highlight those boundaries. The private corporate spaces are visible yet are inaccessible to outsiders who would attempt to encroach into the space. This method of demarcation and exclusion asserts settler colonial-capitalist claims to the land as part of how, as Moreton-Robinson explains, "patriarchal white sovereignty, as a regime of power, operates ideologically, materially, and discursively to reproduce and maintain its investment in the nation as a white possession through the discourse of security."[46] This series of photographs reveals the connection between the surveillance of private property and settler colonial imaginaries of the land as property – a connection that forms the foundational logic of settler states like Canada.

Both *hunting for prey* and *Private Property* creatively contort the tensions between public and private to scrutinize the structures and implications of how these categories are decided and policed, and by whom. Yet, while in *hunting for prey* the viewer is implicated in the piece itself, literally drawn into it as an object of surveillance, *Private Property* keeps the viewer at a distance. Although each work differently frames the relationships between distance and proximity, the boundaries of access, and how the viewer is brought into the piece, they come together in their critique of how settlerhood reproduces colonial territories. Territoriality is a precondition of settler colonialism, and claims to place and land as owned is a central tenant of colonial capitalism in settler colonial states. For Moreton-Robinson, this "white possession" of the land is connected to "white colonial paranoia," or the anxiety of being "dispossessed by racial others."[47] The surveillance apparatus used to secure settler claims to the land is necessitated by the instability of these same colonial claims; the need to constantly monitor and re-establish private property as one's own is what the surveillance apparatus responds to. The two pieces in this section unsettle these colonial-capitalist narratives; specifically, they critically intervene in and visualize the complexities of contemporary surveillance by underlining how surveillance structures operate as an apparatus for settler colonialism's territorial dominance and its possessive and territorial extraction.

Landscape

The three creative projects I discuss in this section deploy recognizable surveillance views of the land-as-landscape to critically engage with historical and ongoing narratives of colonial settlement in the Canadian territory. Often taking the form of a single camera angle fixed on a particular landscape captured over time, these artworks bring to life a contemplative and introspective questioning of the pictured landscape as central to settler colonial imaginary in Canada. These projects reveal how the colonial project renders landscape as either wilderness or settled wilderness: either as a vast and under- or unpopulated landscape available for settlement and extraction or as a cultivated garden-style landscape meant to represent the success of settler dominance. Both approaches to the landscape critically examine national traditions of understanding land as empty and available for settler life.

Depictions of the land as almost or fully uninhabited – such as in the work of the Group of Seven or advertisements for the Canadian Pacific Railway – have been central to narratives of Canadian-ness. The three artworks in this section rethink these narratives, questioning the very idea of landscape and what it reveals of settler colonial assumptions about land being, fundamentally, available. As Carmen Robertson writes, the "reinforcement of empty, rugged landscape in Canada's landscape art tradition erases Indigenous perspectives." Further, she points out that Canada's landscape art tradition "erases ways of knowing land as an interconnected whole where land and thought are one."[48] In other words, representations of empty landscapes have the effect of conceptually framing or parcelling out the land, visually dividing up what is in fact whole, and distancing it from the life already on it. The works in this section draw on viewer expectations of Canadian landscapes as a settler imaginary to ask how land is presented for whom in what way, and to stress how the act of looking is, in itself, a colonial extractive practice.

David Rokeby's installation *Machine for Taking Time* (2001, fig. 3.6), a filmed perspective of the gallery's public exterior space, brings to life questions related to occupying space and time and the construction of perception.[49] Commissioned by the Oakville Galleries in Oakville, Ontario, for the show *Earthly Delights / Deep Gardening*, Rokeby's piece consists of a surveillance camera mounted on a "computer controlled pan/tilt mechanism" that moves through the gallery's garden pathway, taking 1,079 images of this route from specific positions.[50] At each position, the exposure is four seconds. The machine travels the entire pathway over the course of an hour, tracing the same course a visitor would take through this space. The surveillance camera continued to capture this route over a period of months.[51] All of the resulting images were archived, then stitched back together in a video that, while representing that same path, uses images from different days. In the gallery, the video of the collaged surveillance images is projected on a screen that hangs in front of a window through which viewers can see the same path that is shown in the video. The angle of viewing juxtaposes the real-time experience of overlooking the exterior garden space with the virtual experience of the surveillance capture of that same space. While there are occasional glimpses of human and animal life – a person or group of people occupying the pathway, or a gaggle of Canada geese sitting on the grass – the main focus of the video is the manicured landscape outside the gallery. Despite Rokeby's decided centring of the landscape itself, the presence of benches near a small pond, or the appearance of a landscaping truck parked next to the pathway, or the very fact that the pathway is paved, provides clear evidence that the space is curated for human occupation.

3.6 • David Rokeby, *Machine for Taking Time*, 2001.

The camera images curate familiar tropes of viewing the landscape that bring together surveillance monitoring with settler colonial logics of land availability and extraction. The camera position mimics surveillance monitoring by surveying the pathway from a fixed angle; and the camera's framing out of most human activity references a vocabulary of settler landscape art that visualizes the land as empty and available for conquest. The representations of landscape (and landscaping) in *Machine for Taking Time* draw on the aesthetic vocabulary of European and settler landscape art that imagines the land as either vast, uninhabited, and available for settlement, or manicured, civilized, and representative of the cultural taming of the landscape. Both of these conceptions of territory are essentially two sides of the same coin that reflect white supremacist ideals of conquest and occupation. The land as wild and the land as curated are equally constructed visions, and they both replicate an idealized vision of land consistent with the logics of settler colonialism. Rokeby's capturing of the manicured lawns and pathways outside the gallery space illuminates the side of settler imaginaries that encapsulates settler civility and control. The land within the artwork is therefore representative of a

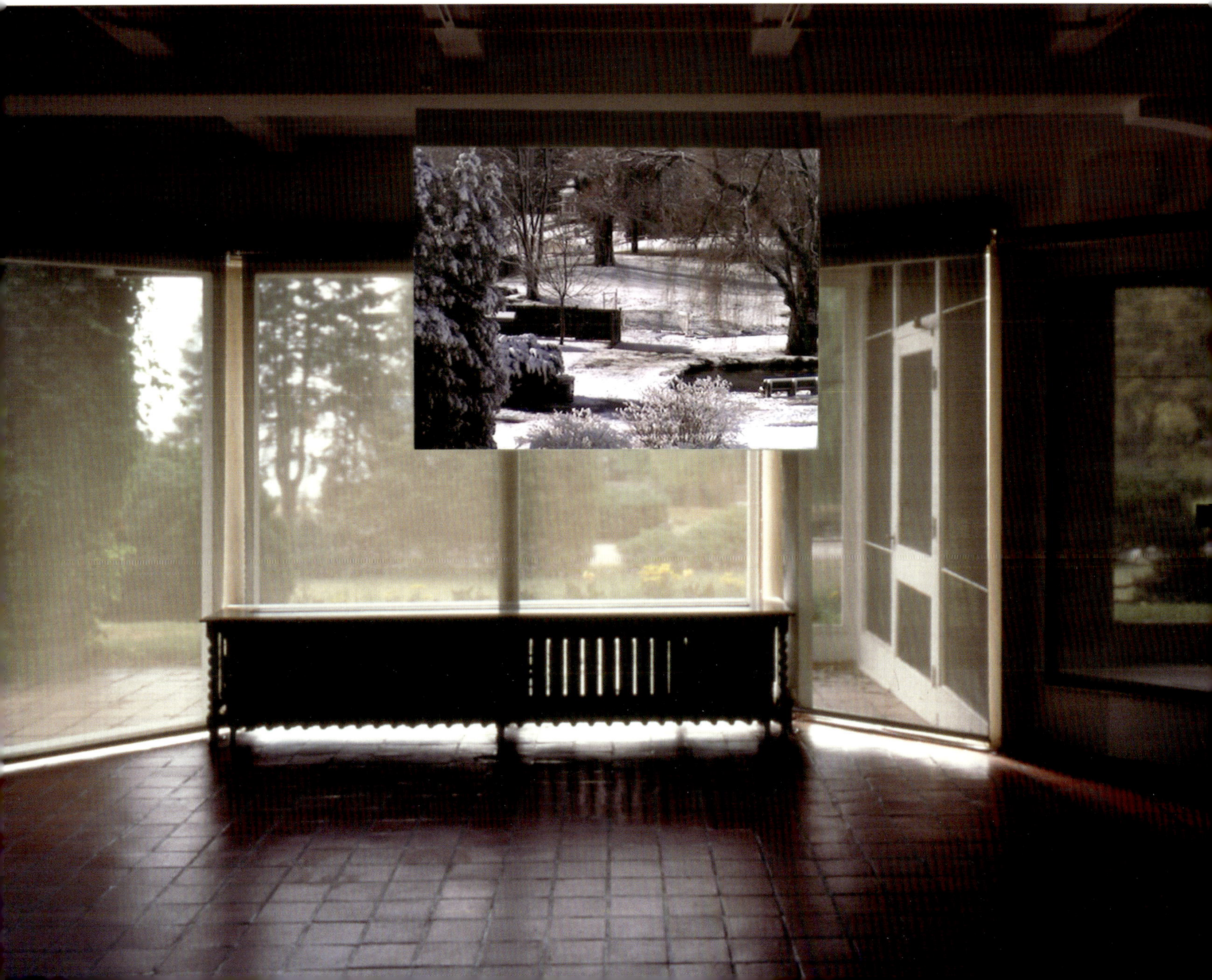

colonial ideal: once land is obtained through conquest, it must be manufactured into an altered state, like a garden, as a form of proving mastery of the land and capacity to structure it toward one's ideal use.

Machine for Taking Time assigns the surveillance gaze as the way to view the land represented in the piece. In doing so, the artwork recognizes surveillance as essential to the colonial project of mapping and claiming the land. Images like these demonstrate the historical continuities of this type of colonial viewing and the surveillance gaze, yet Rokeby disrupts the familiarity of these types of visualizations: by stitching together views of the route from different days and seasons, he refuses the ease of viewing. The artist's emphasis on surveillance cameras to capture and move along this sample of settler-curated space points to the surveillance operations that have enabled settler manipulation and inhabitation of the land more broadly. It also questions how colonial and surveillance "seeings" come together to define, protect, establish, and naturalize settler presence and identity on the territory. This piece highlights the way in which settler logics ensure that the land itself must be policed and monitored to maintain and reproduce settler presence.

As its title suggests, this artwork deals with time not simply as an unfolding of chronological events or images but as a reconstruction of moments, memories, and illusions. While the title of the work, *Machine for Taking Time*, suggests an intended focus on the passage of time, through the stitching together of a collage of images of the pathway taken over a period of time and through different seasons, this work also reveals a relationship to surveillance vision and the operations of public space. The time the machine captures is that of the land marked and clocked by the changing of the seasons. In her discussion of settler time as a concept that informs colonial dispossession of Indigenous people of land and history, Mohawk scholar and political anthropologist Audra Simpson notes the connection of "time coupled with land – to be taken, to be governed and perhaps taken and given back as well."[52] Rokeby exposes the viewer to multiple points in time simultaneously: first, through the video's collaged images, which interconnect different seasons in the visual narrative; and, second, through the juxtaposition of the video screen with the real-time view through the gallery window. These varying vantage points from which to apprehend the landscape repeatedly reposition the gallery visitor in their engagement with the work. Here, Rokeby questions the position from which narratives or perspectives are constructed. Identified as "temporal conflation" by art critic and curator Jessica Fung in her exhibition review, this refusal of linear, passive viewing serves to interrupt "the organization of history around the coordinates of settler occupation."[53] Using the surveillance camera to surveil/survey the land of the gallery, this creative project emphasizes the role of surveillance in constructing narratives of land and place, as well as the ways in which these narratives can be shaped and moulded to a particular way of seeing. *Machine for Taking Time* foregrounds the ways in which time is an often unseen or invisible element in how understandings of place, land, and belonging are produced.

Another work that uses temporality to contemplate the role of surveillance as a mode of understanding place in the Canadian context is *Meta Incognita* (2010, fig. 3.7). Directly referencing Rokeby's *Machine for Taking Time*,[54] this work is a time-lapsed video that consists of over five hundred images taken from a public weather webcam located in Kimmirut, Nunavut, on Baffin Island. For this work, artist Pierre Tremblay began archiving the weather camera's images on 21 June 2010 – the longest day of the year – archiving images every fifteen minutes, twenty-four hours a day. Each image contains a representation of the land captured from the static perspective of the weather camera, depicting a series of houses lining the sides of a small road, with a street lamp and wires cutting through the middle of the frame. Distant mountains and a large body of water fill the background space, providing a clear picture that this is a community on the edge of a vast landscape. Some of the images also contain text that includes weather information and a time and date stamp, which narrates and describes the larger context of each scene. *Meta Incognita* takes these archived images of the land captured automatically by the webcam and recontextualizes them into a video that gives motion and movement to the weather cycles in the images. This creative work has had multiple variations and is part of a larger collaborative project between Tremblay, David Bouchard, and Bruno Lessard, with audio composer Alex Geddie, entitled *Nunavut Lights*, which explores different ways of representing and visualizing the

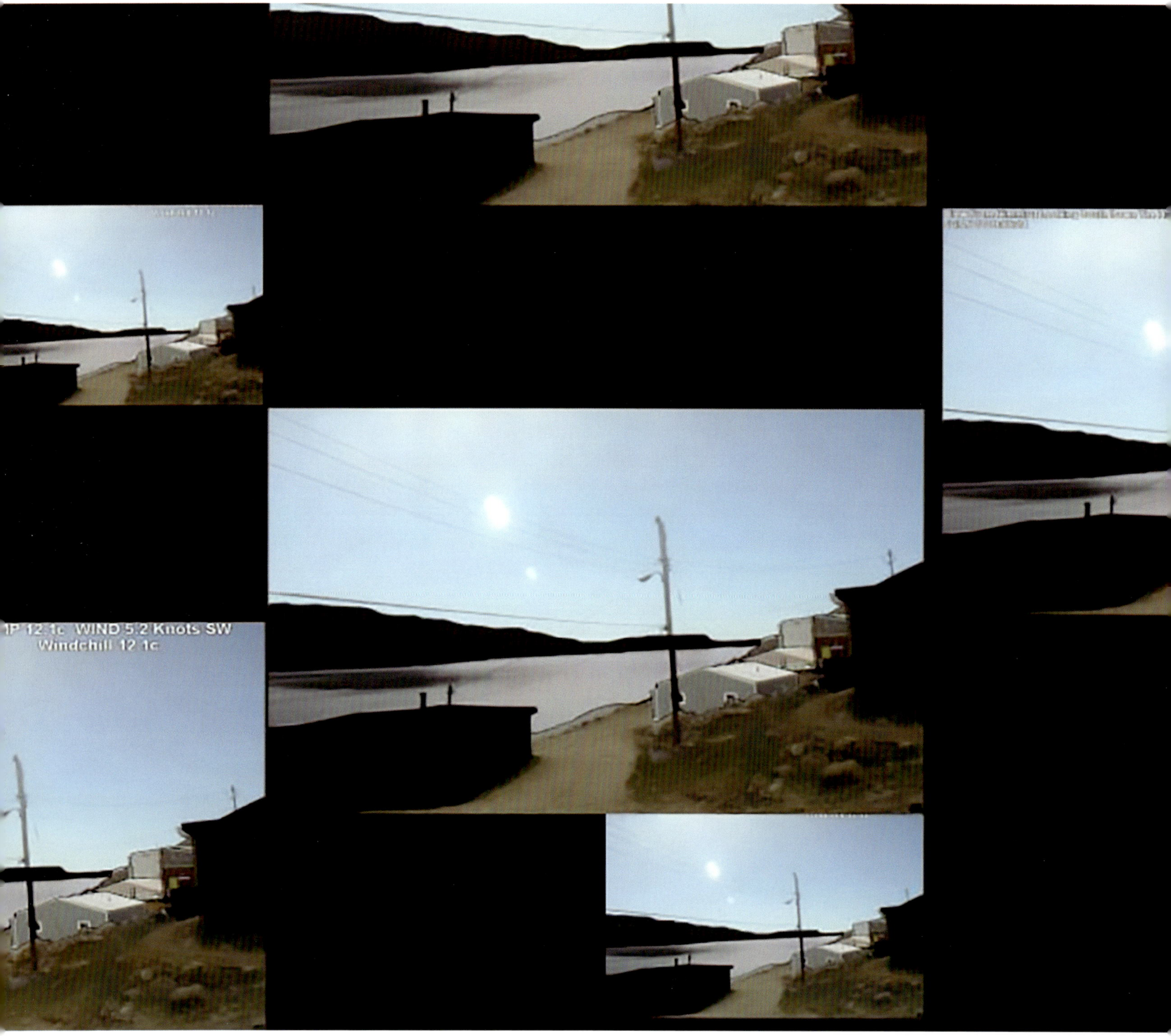
P 12.1c WIND 5.2 Knots SW
Windchill 12.1c

3.7 • Pierre Tremblay, *Meta Incognita: Summer Variations*, 2010.

photographic archive. Tremblay, Bouchard, Lessard, and Geddie mounted the work of this collaborative project in an exhibition entitled *Constructed Land* (2012).

According to Tremblay, the artwork uses the weather webcam to raise "questions about representations of time, the photographic archive, surveillance and the documentary genre."[55] On first inspection, using a weather camera might seem unexpected in the context of surveillance perspectives, considering that it is not conventionally associated with overt forms of surveillant policing technologies. A deeper look, however, reveals its role in the more mundane, familiar, and omnipresent forms of surveillance logic that situates certain types of looking, seeing, and knowing as harmless but ubiquitous. The weather camera is publicly accessible, providing pertinent information for locals to help determine their activities for the day, as well as constant insight into the region for those who are intrigued from a distance. By bringing the weather camera into the art context, emphasizing his manipulation of the images and archive, and thereby forcing a different type of engagement with the representations, Tremblay challenges the idea that the lens of a camera is ever unbiased or objective and that the archive is never unmanipulated. In his digital manipulation of the images he has collected and archived as part of this project, Tremblay redefines the language and frame of the camera. Certainly, through his reappropriation of the weather camera's online stream, he reveals the insidious forms of looking in the everyday and the logic of surveillance that infiltrates unexpected and presumably mundane contexts.

This logic of surveillance is found not only in the *way* something is filmed – that is, by replicating certain recognizable angles of viewing – but also in *what* is filmed and for whom. When considering surveillance and art in their intersection with public space, it is important to think about which landscapes are available for consumption, access, and extraction, and which are not. *Meta Incognita* highlights the disconnect between the viewer and what is being viewed, enacting what Bouchard, one of the other artists involved in the *Nunavut Lights* project, describes as "a kind of virtual tourism."[56] To be sure, the land as it is represented in the artwork (as well as the original output from the weather camera itself) is constructed and presented from the

3.8 • Philip Cheung, *Ranger ATV Patrol, Taloyoak*, 2017.

perspective of people other than those who live on, inhabit, and occupy the actual land – in this case, the perspective of the artist or artists. In an interview, Bouchard confirms that this type of artwork presents "landscapes which lie in between real and imagined territory."[57] Yet such a statement raises the question, imaginary to whom? As Steven Loft, an art critic and director of Indigenous art programs at the Canada Council for the Arts, writes in his exhibition essay that accompanies an incarnation of the *Meta Incognita* piece, the fascination with northern landscapes is a long-standing subject of settler colonial imaginaries, and the region of Nunavut "is also the frontier of resource extraction, national 'sovereignty' and colonial encroachment. It is, indeed, a land of ancient tradition and modern settler politics in an often uneasy and contentious relationship."[58] *Meta Incognita* amplifies the ways in which surveillance cameras can manifest a form of distanced viewing that prioritizes the knowledge of the viewer of the land over the knowledge and lived experience of the people on that land. By framing access to the Arctic through remote cameras, the work critiques how such visual systems reproduce settler colonial power dynamics, turning place into spectacle and knowledge into something to be extracted.

A photographic series by Philip Cheung further brings to light the tensions around landscapes that capture issues of sovereignty, settler extractions, and state making. Cheung created his series *Arctic Front* (2019) during his time in the Canadian Forces Artist Program, during which he was embedded with the Canadian Rangers, a unit of part-time reservists of the Canadian Army Reserve composed primarily of Indigenous members.[59] The photographs in *Arctic Front* are quiet, contemplative portraits of ongoing Canadian military presence in the Arctic. Exhibited at Circuit Gallery in Toronto, from 10 January to 2 February 2019, the photographs in the series depict open scenes of rocky landscapes, dense bogs, jagged shores, and expansive seascapes, with individuals or groups of people wearing bright red sweatshirts or camouflaged fatigues appearing throughout, marking the land, sea, and sky with their presence. Depicting rangers variously camped out on a rocky beach, dispersed across a peninsula on ATV (all-terrain vehicle) patrol (see *Ranger ATV Patrol, Taloyoak*, 2017, fig. 3.8), or otherwise congregated on the land examining the foliage, the images rehearse rituals

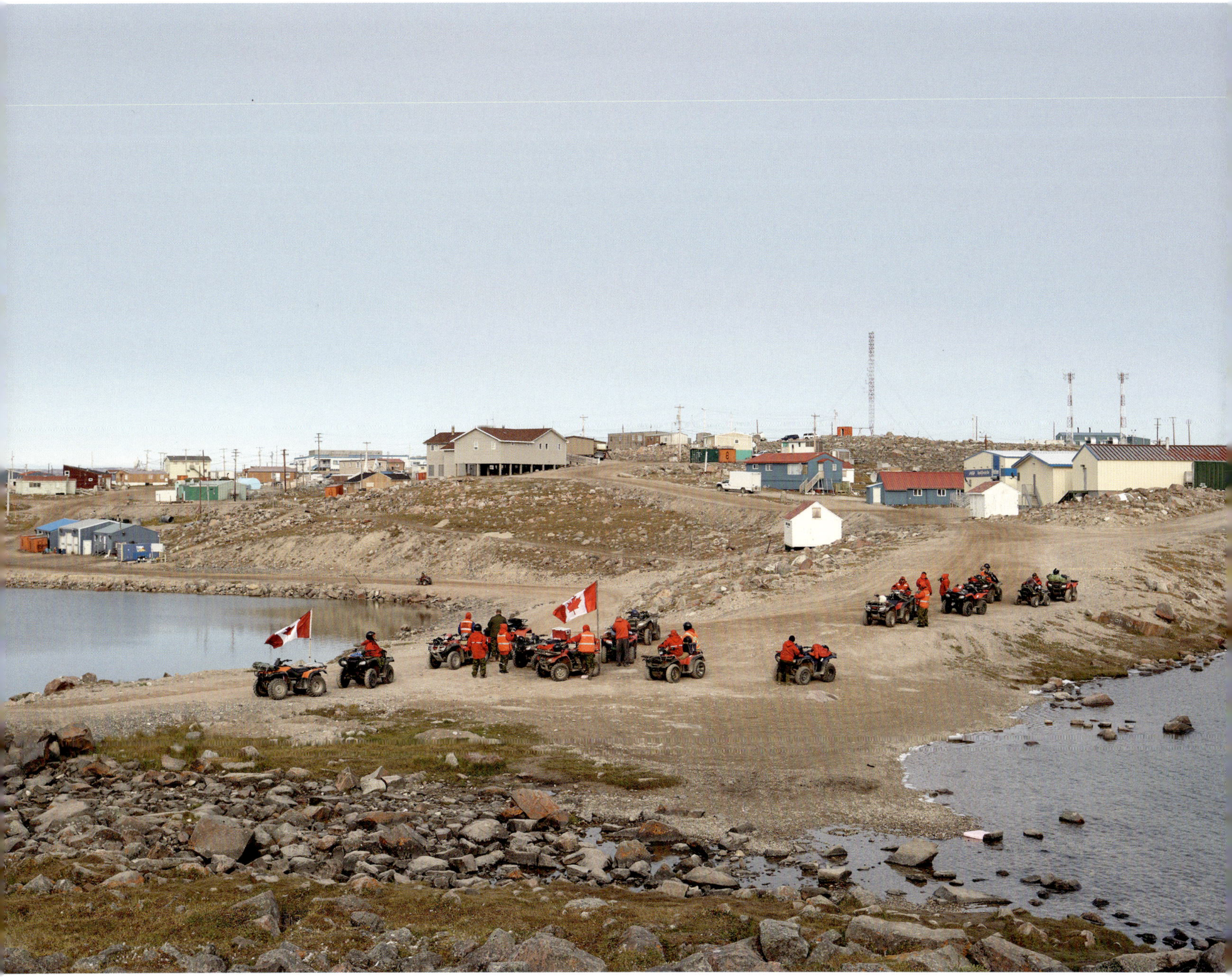

of surveillance – of watching, of monitoring – as part of the ongoing maintenance of Canadian state making and claims to the land. Cheung's series brings forward questions about the role of surveillance logic in ongoing debates around sovereignty and colonial apparatus. The series presents the colonial encounter of Canadian state actors who are maintaining occupation and vigilance on national territories, yet it complicates this narrative through the identities of the Canadian Rangers, who are primarily Inuit, First Nations, and Métis people from the local communities.[60] As the gallery's exhibition description explains, *Arctic Front* presents "subtle, indirect portraits of the dissonant tensions in the North between military and civilian (both settler and Indigenous) interests, resources, and labour and subsequently a contemplation of the human capacity for adaptation, survival, and endurance."[61]

In this series, surveillance logic operates as the foundational presumption of how knowledge, control, and power are deployed in colonial-dominant contexts. The act of looking becomes a form of knowing, authority, power, and control; to monitor and to observe is to ensure that the space of the land and sea are protected from encroachments that would challenge Canada's claim to the space and to the resources it holds. The photographs represent some of the activities of the Canadian Rangers, who are directed by the Department of National Defence to provide "support to sovereignty operations such as reporting suspicious and unusual activities as well as collecting local information of military significance."[62] The Arctic holds a specific place within the Canadian settler colonial imaginary, sparsely settled and with a climate and elements that are considered harsh to the foreigners who occupy and make claims to it. Certainly, the Arctic is an intensified military frontier, as nations such as Canada, Russia, the United States, Denmark, and Norway attempt to settle disputes over new avenues for shipping and opportunities for oil and gas extraction that the ice melt associated with climate change has opened up.[63] Very few, if any, of these debates acknowledge the continued presence of Inuit communities in this region, who themselves make claims to sovereignty on this land.

Cheung's photographic series makes a subtle yet significant contribution to unpacking the tensions and ongoing colonial encounters of this region as a military front, a context that Cheung points to in the title of the series. *Arctic Front* is a slow meditation on the landscape and on the relations that mark its land and history. Compositionally, the photographs are not unlike those survey photographs that (in)famously charted and mapped earlier Canadian expansion. Cheung's photographs critically intervene into these problematic visual histories by bringing these relationships to light and, as communications studies scholar Sara Matthews points out, by questioning how "resource extraction, claims for territory and rights of passage, environmental conservation, tourism, scientific exploration as well as the manufacture of the North as somehow essentially 'Canadian,' make up these settler relations which litter the landscape with the physical ghosts of their infrastructures."[64] The series spotlights the colonial presence of Canadian state actors in maintaining occupation of and vigilance over national territories, and the importance of ongoing surveillance structures to these processes.

The angle of Cheung's photographs in *Arctic Front* differ from that in *Machine for Taking Time* or *Meta Incognita*: while the latter two works incorporate angles of looking that immediately recall a surveillance lens – angled, top-down viewing – Cheung's photographs demand a closer look to pull out the surveillance threads that bind all three of these projects together. Despite this difference, Cheung's work joins the others in representing land (and sea and sky) as space to be monitored, tracked, surveyed, and surveilled, with the intention to know and control. The distance between the viewer and the spaces differ: both *Arctic Front* and *Meta Incognita* imagine the North for viewers geographically distanced from that locale, while *Machine for Taking Time* intimately ties the act of viewing the land with inhabiting that very space, although separated by time and by interrupted continuity. Yet together, all three works make conspicuous how looking itself can be an extractive process tethered to settler colonial visions and expectations. These three artworks present a sense of landscape as critically tied to settler colonialism and surveillance viewing, and work to disrupt and call into question the ways in which these forms of looking are utilized as part of an ongoing settler colonial project in Canada.

Surveillance visions of the land have been and continue to be integral to the settler colonial project. In critically drawing on these visions, the eight artworks in this chapter show how, in surveillance sociologist Midori Ogasawara's words, "nation building and colonialism are intertwined in the development of surveillance systems" within the Canadian context.[65] By foregrounding the inextricable relationship between settler colonialism and surveillance, the artworks each denormalize that relationship, challenge settler claims to the land, and disrupt the role of land within the settler colonial project. While operating in different ways and using various mediums, these artworks come together to interrogate the construction and maintenance of settler visions by engaging with colonial perceptions of and relationships to land and territory, and surveillance's role in these processes. The utility of these works, and my analysis of them, is not to centre or re-emphasize settler colonial histories in discussions of the land but, rather, to dismantle and decolonize these histories through challenging their normalization.

Each section in this chapter takes the larger relationship of surveillance with the land and breaks it into different approaches. In "Mapping," the artworks problematize mapping and cartography as visualizations of settler imaginaries of territory; the creative projects in "Private Property" ask how surveillance monitoring reproduces settler colonial systems, via property ownership; and in "Landscape," the works question how settler dominance manifests in renderings that picture the land as either uninhabited wilderness and available for settlement, or settled wilderness and representative of settler dominance over the land. The production and circulation of these artworks in the post-9/11 period, defined by the dominance of the policies and rhetoric of the War on Terror, make present the continuities of historical and ongoing projects of settler colonialism. They also elucidate how expansive projects of security and surveillance work in reciprocity with ongoing histories of colonialist empire building, racism, and white supremacy. These art projects – and my engagement with them – articulate the imperial logic at the heart of contemporary surveillance structures. In so doing, the works demonstrate the operations of surveillance as a vehicle for naturalizing settler colonial life on the land, produce new knowledges of surveillance and colonial attachments to land, and engage viewers in unique ways to rethink their own understanding of these ongoing histories. All the works in this chapter demonstrate that surveillance viewing is settler colonial viewing.

4 Resisting

Our charge is to make imagining liberation under oppression completely unthinkable, to really push ourselves to think beyond the normal in order for us to be able to address the root causes of people's suffering. That's the politics we should be focused on.

Mariame Kaba, *We Do This*

The artworks in this chapter are united by a strategy to use art as a medium to invite in and engage audiences as informed communities. This strategy is born out of a desire to push back on and confront surveillance, both to challenge it as a form of social injustice and to imagine alternatives to surveillance and its logic. By centring such a strategy, this chapter differs in its focus from the other thematic chapters in this book, each of which examines a different dimension of surveillance logic: the surveillance gaze, surveillant policing, and surveillance visions of the land. By contrast, in this chapter, the creative projects come together in a desire to challenge both surveillance logic and, as I argue, the way in which this logic bolsters settler colonialism; by resisting one, they resist both as mutually constitutive structures. They strategize to rethink and reimagine how surveillance viewing is settler colonial viewing. In so doing, they work to educate publics, as part of a strategy to present surveillance logic and its applications as a social ill and, as activist Mariame Kaba writes in the epigraph to this chapter, "to think beyond" current systems of oppression and authority. They share an intention to look back at, evaluate, and resist surveillance as a social injustice and to position art as a medium through which larger audiences can be engaged in unique ways.

The artworks I have organized into thematic chapters could each fall into multiple categories. The nine works that appear in this chapter could be discussed with respect to the other chapter themes; by the same token, some of the other artworks in this book could be interpreted as sharing the motivations I discuss here. I have selected the nine works for this chapter because they share in helping bring to light new ways of thinking about creative practice as a medium for being in and engaging with the world. To my mind, they function beyond bringing awareness to the contours and contestations of surveillance logic: they present ways of understanding that challenge past, present, and future conceptions of Canadian settler colonial life where surveillance is omnipresent and inevitable. Additionally, they work to bring art audiences together as a curious and engaged group, in a political desire to dismantle surveillance's alienating design that disconnects people from the safety and care of community.

I think of these efforts as strategies of resistance, ways in which to challenge dominant paradigms of surveillance logic and to offer alternative models of being in the world. In the introduction, I discussed the visuality of surveillance and the role of art in revealing and shaping it. To build on that framework, in this chapter I examine how selected artworks resist the authoritative visualities of surveillance logic in its relationship to settler colonialism and other systems of oppression, such as racism, white supremacy, ableism, and cisheteropatriarchy. These creative projects create a path to claim the "right to look" and to challenge the normalized authority of surveillance logic in the contemporary Canadian context. For visual culture theorist Nicholas Mirzoeff, the right to look is not simply an aesthetic challenge to the expectations of what a visuality – such as that of surveillance – can show, but rather "the grounds on which such assemblages can register as meaningful renditions of a given event."[1] The nine creative works in this chapter resist surveillance logic not just because they challenge and reframe the way surveillance sees and what it shows, but also because they counter the premises, beliefs, and expectations on which that seeing and showing is built. They highlight particular elements – such as care, being in relation, collective actions – that are antithetical to the individuating and dehumanizing logic of surveillance to show its limits and weaknesses and to resist its encompassing gaze. Resistance here offers possibilities to dismantle surveillance logic as a vehicle of settler colonialism and to create meaningful alternative ways of being in and understanding the world within a contemporary Canadian context.

The artworks included in this chapter can be seen in the larger context of creative strategies of resistance – as checks and balances that hold power to account by turning the camera back on surveillance authorities. Activist-minded groups have utilized resistance strategies in their own challenges against state or corporate surveillance. Groups like El Grito de Sunset Park and Copwatch Patrol Unit in New York City integrate traditional forms of grassroots activism with digital technology to challenge over-policing, gentrification, and lack of self-determination in communities of colour.[2] Drone Warriors, an Indigenous-led group that grew out of the Water Protectors standing in opposition to the proposed Dakota Access Pipeline, use photographic drones to monitor the human rights and constitutional violations enacted by a militarized police force and to document the serene landscapes that the proposed pipeline threatens to disrupt.[3] The works I include in this chapter are in conversation with these international art-activist activities. They deploy resistance as method, not simply because they "look back at" at authoritative surveillance in the literal sense but also because they reframe and reimagine surveillance visions to resist and move beyond structures of settler colonialism and its associated systems of oppression. By positioning the artworks as sharing in this critical activist ethic, I also make a claim that these works have an affinity with traditions of art activism that situate art as a social and political practice, using creative projects and representations to push back and challenge systems of domination, authority, and inequity.

Art and activism have long been intertwined, with artists challenging societal norms, provoking critical thought, and advocating for social change through creative expression. Artist and activist Gregory Sholette describes activist artists as having a "prevailing desire ... to transform their practices into a form of highly focused protest."[4] While I argue that the works included in this chapter fit within categories of resistance and art-as-activism, I do not presume that the artists all intend explicitly to be activists. Rather, I see the creative practices themselves as challenging and resisting surveillance systems and as working in conversation with activist initiatives to refuse futures in which the proliferation of surveillance in the defence and maintenance of settler colonialism is inevitable. Art historian Kirsty Robertson reminds us that, while "not all activists are artists, and certainly not all artists are activists," there is still great "potential for art as a radical gesture or intervention."[5] This potential in these works is rooted in their social engagement – the ways in which they bring audiences into critical contexts to look back at, scrutinize, and assess surveillance as a social injustice. The humanity of the audience is crucial here, in contrast to surveillance's tendency to dehumanize people as data aggregates, foreclosing the possibilities of community and care. The works in this chapter are, in the words of urban design scholar Quentin Stevens and geographer Julia Lossau writing about art in public spaces, "invested with a social or communal focus that include processes of communication with the public."[6] These works create communities from art audiences by bringing people into a shared critical and creative resistance framework, as an effort to challenge the larger structures that inform surveillance logic in the settler colonial context of Canada.

Consistent with the critical role of art as a vehicle to think through, trouble, challenge, and resist authoritarian contexts, creative projects in Canada that address surveillance systems increased in number following the start of the War on Terror. Artists and cultural practitioners increasingly engaged with the systems that justified elevated and widespread surveillance measures as part of a public reckoning. Since the start of the War on Terror – with its logic of increased surveillance practices for the sake of national and global security – Canadian domestic and international surveillance policies have persistently expanded, generating controversies over the government's capacity for broad (and often unknown and non-consensual) access into the lives and activities of private citizens.[7] In this context, surveillance systems are often framed in terms of seeing, uncovering, and revealing. But importantly, I argue, surveillance logic is also about not seeing the fabrication of the settler state. The logic of surveillance helps to make imperceptible the mechanisms that create and maintain the settler colonial state, mechanisms that involve the policing and violent displacement and genocide of Indigenous people as the original inhabitants of the land, and ongoing systems of violence and oppression that continue to inequitably target Indigenous, Black, racialized diasporic, 2SLGBTQIA+, and disabled people and communities. These systems enact what Audra Simpson identifies as settler colonialism's desire "to destroy what is not. The state does so with a death drive to eliminate, contain, hide and in other ways 'disappear' what fundamentally challenges its legitimacy."[8] In this, the state wields surveillance as a key tool.

The artworks here refuse this "disappearing" action of settler colonial contexts by presenting a lens onto processes of surveillance logic. They bring forward questions and concerns regarding surveillance technologies and the often unseen and mundane belief systems that guide them. As I explain in other chapters in this book, surveillance – as a form of settler colonial looking – evaluates, produces, claims, and protects settler territories and the extractive activities on them. Importantly, it does so in a way that normalizes and superficially depoliticizes these activities, asserting forms of looking as a necessary and inevitable part of securing the nation. In this context, surveillance structures come to be framed as neutral or impersonal. As feminist scholar Rosalind Gill articulates regarding the widespread and sometimes uncritical acceptance of surveillance activities by the public, "governing thus becomes recast as a technical rather than political activity."[9] The artworks here refute the apolitical cast of surveillance. They participate within what Jennifer Gradecki and Derek Curry identify as "a socio-technical assemblage of new media and social networks" that work to rethink, trouble, resist, and imagine beyond the existing structures of surveillance that uphold systems of inequity and oppression in settler colonial Canada.[10]

As facets of this "socio-technical assemblage," the artworks included in this chapter find innovative ways to confront and expose the surveillance apparatus, questioning its ethics, impact on personal freedoms, and biased applications. They do not simply inform art audiences about the dangers of contemporary surveillance logic and its operations, however; they also create engaged communities out of art audiences and imagine other worlds where state surveillance mechanisms as a form of expansive control are not inevitable. As part of their broadly envisioned method of resistance, they work within a politics of *being in relation*, of bringing people together to resist the dehumanizing and disconnecting glare of surveillance logic in the context of settler colonialism. The nine works explored in this chapter are divided into three sections – "Belonging," "Intimacy," and "Community" – which provide different angles of address on this topic. In "Belonging," the artworks interrogate the bureaucratic markers that delimit movement, access, and thus belonging and inclusion within the Canadian context; in "Intimacy," the creative practices examine issues of vulnerabilities and closeness as a limit to the surveillance gaze; and in "Community," they offer alternative ways of seeing state and corporate surveillance tactics to create a collective space of shared social and political engagement. The creative projects here utilize surveillance resistances as a strategy to trouble mainstream conceptions of surveillance. Yet, they are not simply "looking back" in the literal sense of flipping the surveillant gaze; rather, they recalibrate the surveillance lens for the express purpose of raising awareness around the logics of surveillance as upholding the tenets of settler colonialism – bolstered by systems of white supremacy, racism, ableism, and cisheteropatriarchy – as a clear social problem. It is from this foundation that their artistic strategies and (intentional or not) activist ethics flow.

Belonging

The three artworks in this section question how surveillance systems inform the politics of access and belonging to, and inclusion in, the nation-state of Canada. I use them to build on the discussion of surveillant policing arising from the creative projects in chapter 2 and to consider policing activities with respect to settler colonial histories in Canada. Positioning the passport as a surveillant policing marker of mobility and inclusion, these art projects assert that strategies of surveillance inequitably assess who is considered intrinsic to Canadian nationhood and who is deemed as external and a potential danger to the state. They use or implicate the bodies and movement of members of the art audience to directly engage them in questioning their own relationship to belonging and how this relationship may uphold the exclusion of others. These works ask questions such as: Who is selectively included as part of the cultural and political fabric of Canadian nationhood? And how does the state use surveillance and security structures as policing agents to protect and safeguard the nation and its attendant imaginaries on behalf of this assumed citizenry? These creative projects work in conversation with critics of citizenship and nationalism such as Robyn Maynard, who writes, "citizenship, as it exists today, remains a legitimate, widely and formally accepted form of exclusion, allotting rights and protections to some and withholding it from others who then fall victim to state violence by virtue of their designation as 'outsiders.'"[11] Surveillance systems inside and outside the borders of the nation-state reinforce citizenship as a basis for exclusion in its claim to protect against outsiders who would challenge or disrupt the safety and sovereignty of the nation.

For sociologist Nandita Sharma, racism and nationalism are co-constituted: belonging to the nation is premised on a certain racialized expectation of which inhabitants compose the central body of the nation-state. Within Canada – which professes a commitment to multiculturalism and thus officially recognizes multiple racialized groups within its citizenry – foundational national tenets of white supremacism along with rising Islamophobic, anti-Black, anti-Brown, and anti-Indigenous sentiments in the post-2001 era hold different groups of people as more Canadian – as more included in the fabric of Canada – than others. As Sharma writes, "racist definitions of national belonging result not only in the refusal to admit certain non-citizens but also in the social refusal to accept many *co-citizens* in the political community. In a very real sense, negatively racialized citizens do not enjoy the same *de facto* rights as those with the unqualified assurance of national belonging."[12] The creative works in this section unpack different iterations of such qualified belonging by examining the passport as a bureaucratic accessory of surveillance logic that acts as an official endorsement of national belonging, both at the nation's borders and within them. Each creates its own meaning but also works in dialogue with the others, as they share a motivation to reimagine and reinvent alternative forms of togetherness and belonging in those spaces that often work to malign, segment, and exclude. Positioning the passport as the marker of mobility and inclusion, these three art projects think through ways of redefining forms of belonging to oneself and each other that move beyond these exclusionary systems.

Ayesha Hameed and Anita Schoepp's performance piece *Papers Please!* (2008, fig. 4.1) uses surveillant policing tactics to confront individuals in a public space and to pose questions about whose presence there is expected and, thus, politically invisible. In this work, Hameed and Schoepp, who are wearing official-looking uniforms and hats, cover their faces with white paper "smiley" masks and position themselves as good cop / bad cop on a document-checking spree. The good cop, named Forgetting, carries an oversized packet of tickets and a pencil, and cheerily writes citations for shoppers in the area, as well as for inanimate objects such as plants and cars. Remembering, the bad cop, with an obviously fake security camera mounted on their shoulder, scans all people walking by. Hameed and Schoepp imperfectly replicate recognizable surveillant policing protocols to confront audiences with their own expectations of who these structures serve within Canadian public life. *Papers Please!* is the first of three parts in their series entitled *Borders in the City*, which uses technologically simple materials – such as paper, chalk, and radios – to, in the artists' words, "explore how invisible borders affected the construction of public memory and the circumscription of bodies deemed undesirable by the state. These interventions took place over eight months and sought to question the technologies of state control and surveillance, and the practices of exclusion and inclusion that are complicit with such technologies."[13]

In the first part of the series, the two "cops" walk along the street arm-in-arm, shouting "Papers please!" This activity invokes "the itinerant nature of border control in the form of racial profiling and the targeting of 'undesirables' on the street by the police."[14] The people these cops target are not the usual suspects, however.

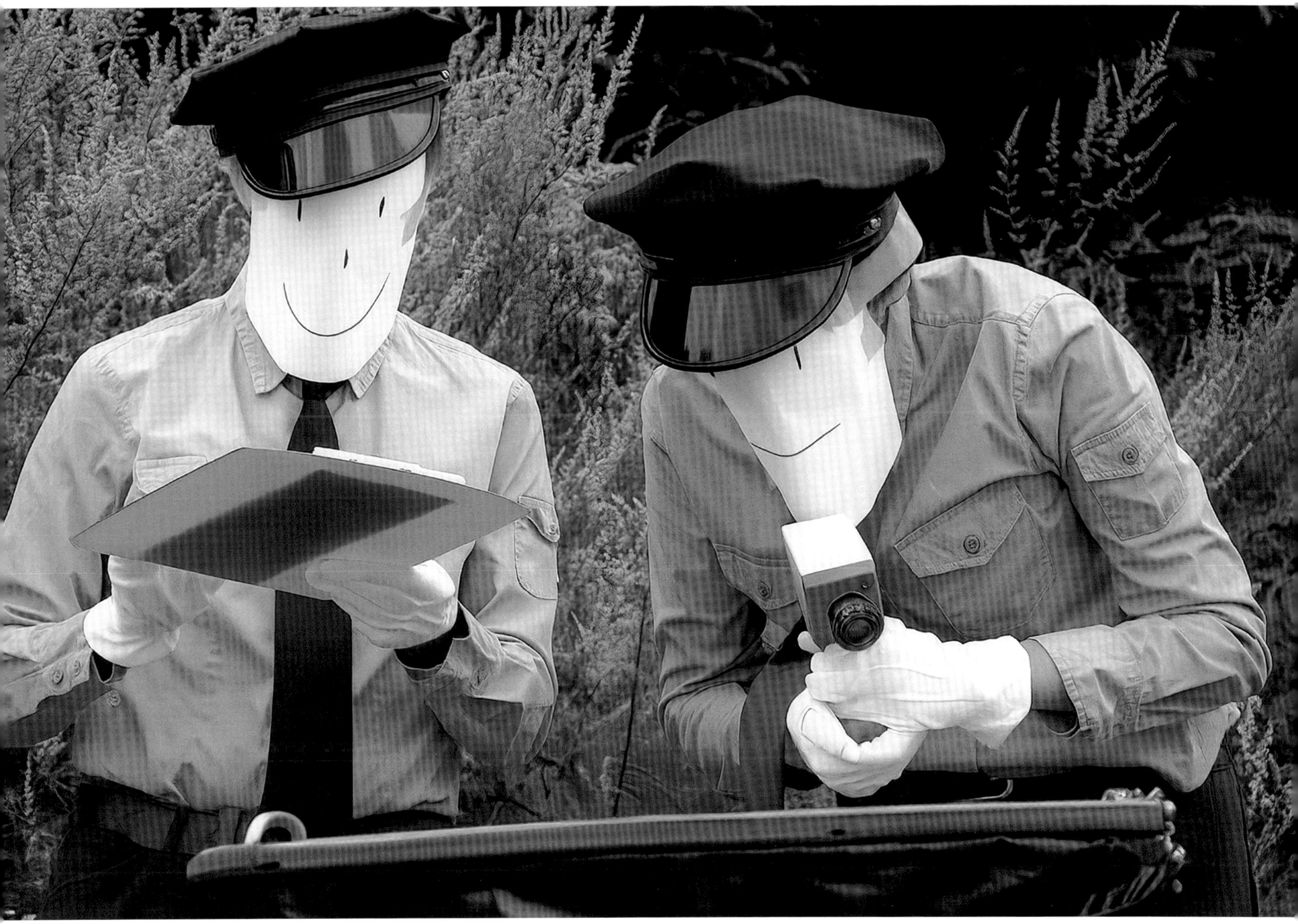

4.1 • Ayesha Hameed and Anita Schoepp, *Papers Please!*, 2008.

Those most often caught in the crosshairs of surveillant policing measures – such as youth, racialized people, or unhoused people – are not the focus of the duo's investigation. Rather, the process of interpellating who is considered a security risk within the expectations of the Canadian settler colonial nation-state is reversed, because Remembering and Forgetting approach, as Hameed describes, "wealthy white people on the street."[15] The piece flips the script, so to speak, on who becomes the object of surveillance. The security officers push their own logic of criminal profiling even further by ticketing inanimate objects as well. In circulating these tickets to groups of people (or objects) usually outside the focus of such surveillant policing apparatuses, this performance makes visible the daily experiences of targeted groups. They make them visible by literally acting out the often unseen bureaucratic exchanges that targeted groups must endure. Hameed explains: "In this performance, Remembering and Forgetting embody the technologies that enforce borders and segregation … with all the stupid, mindless and Kafkaesque bureaucracies of such technologies of control. This type of border control is enforced simultaneously by agents of the state, in the form of military personnel, technologies of control, surveillance cameras, tickets, fines, and threats of detention and incarceration."[16] In their embodiment of such technologies, these characters move about, asking passers-by for papers and handing out tickets on the street that read "Description of Offence: Accommodating Reasonably. Description: Not having proper papers. In addition, we are offended because you are in the wrong place at the wrong time, you are the wrong colour, religion, and you're just wrong for Canada.'"[17]

One of the offences listed in the ticket, "Accommodating Reasonably," references Quebec's Consultation Commission on Accommodation Practices Related to Cultural Differences, which sought public feedback on the accommodation of religious and cultural practices of racialized diasporic people and communities in the province in 2007. The so-called reasonable accommodation debates and resulting provincial commission (dubbed the Bouchard-Taylor Commission after its co-chairs) were a response to heightened public tensions surrounding a series of controversial cases that grappled with the extent to which immigrant and minority religious practices ought to be accommodated in Quebec.[18] These cases included a soccer referee forbidding an eleven-year-old girl from wearing her hijab during a game, and a school prohibiting a Sikh boy from wearing his ceremonial dagger. One of the more public cases was that of the small, predominantly francophone town of Hérouxville adopting "a 'life standards' code of conduct for immigrants," which, as feminist and cultural studies scholar Gada Mahrouse explains, banned "its residents from practising female circumcision, stoning and immolation – all practices associated with a perception of Muslim barbarianism [despite] the fact that Hérouxville's population is almost entirely French-Canadian, with few, if any, Muslims."[19] Growing out of these types of controversies, the reasonable accommodation debates set the ethnic Québécois identity – that of the historically white, Catholic, French-Canadian majority population – in opposition to the needs and practices of ethno-cultural and religious minority groups such as Muslim, Sikh, and Jewish peoples.

The basic rationale underpinning these debates was rooted in deep-seated racist, anti-Black and anti-Brown, antisemitic, and Islamophobic belief systems. It brings to mind Jasmin Zine's assertion that "the idea of policing the boundaries of belonging and banishment within the nation out of a fear that foreign outsiders are taking over and undermining the way of life of 'old-stock Canadians' is a common sentiment and basis for racist and Islamophobic conspiracy theories among far right, White nationalist groups."[20] Zine's analysis underscores how the anxieties and belief systems that evaluate belonging and threat are not aberrations but structurally embedded within Canadian (and Québécois) identity, making it unsurprising that such ideologies surfaced in these debates and demanded institutional response. The final report of the Bouchard-Taylor Commission highlighted the need for greater efforts to combat racial, cultural, and religious inequalities and discrimination in the social and political life in Quebec.[21]

With *Papers Please!*, the artists reference such discussions and draw them into a new context defining what is in need of accommodation and thus constituted as outside the cultural norm. The performance interrogates the evidence one requires – papers, passports, clothing, skin colour, class markers – to move through society

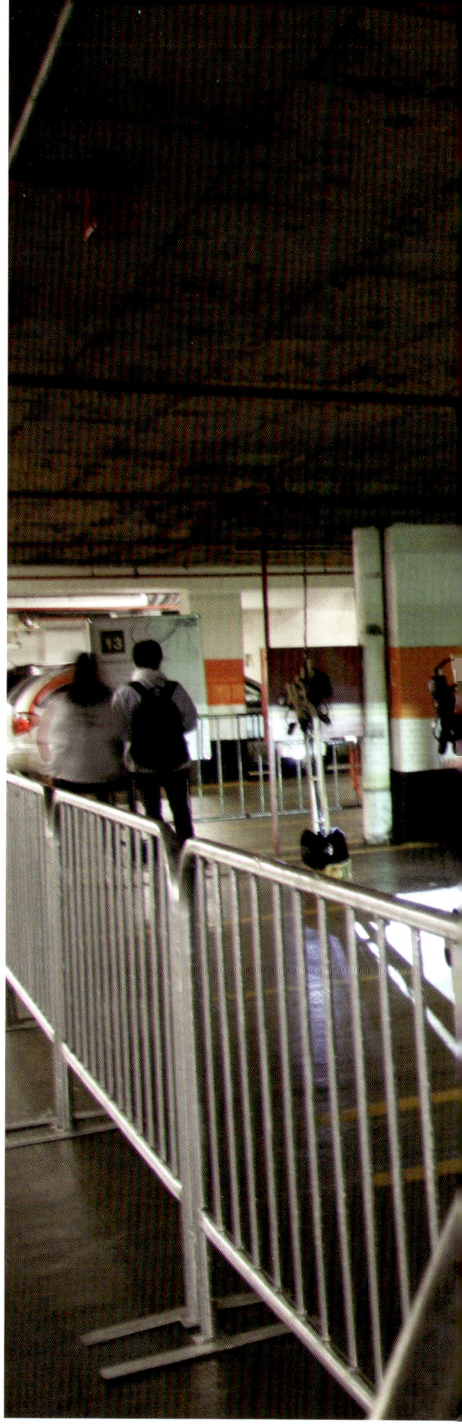

unchecked. By turning the lens of monitoring and surveillance on those people who normally enjoy unqualified access to their rights to exist and engage in their cultural or religious practices, *Papers Please!* forces them to occupy the other side of the so-called debate. This creative intervention sets up confrontations between surveillant policing apparatuses and people who are often exempted from their scrutiny, and compels those people to reckon with what it means to be deemed outside the cultural, religious, and racial norm. The work asks them to reflect on what it might feel like to be singled out as undesirable by the state and therefore inequitably targeted by surveillant policing. In forcing an encounter between the surveillant policing apparatus and those who most often remain outside of its net, *Papers Please!* encourages the public audience to question their relationship to and complicity in systems of exclusion and inclusion that determine the level and frequency of contact with policing agents of the settler colonial state.

Providing a creative way to re-engage with dominant surveillance systems through inviting audiences behind the proverbial surveillance curtain, Hameed and Schoepp's *Papers Please!* draws public attention to the inequities of the surveillance gaze and its potential violences, particularly toward historically over-surveilled people. Drawing public attention to surveillance as a social injustice in this way may feel uncomfortable for some unwitting public participants. However, the broader aim of this piece is to bring people into a common experience, although obviously the material stakes for the artwork are very different than those in everyday life. The next two works in this section, which build on the ideas Hameed and Schoepp introduce here, seek ways to rethink spaces of belonging and togetherness, and to reclaim self and community against and beyond those terms prescribed by the settler colonial state and the surveillance structures that uphold it.

The artist collective Bambitchell developed an art project called *Border Sounds* (2011–15, fig. 4.2), which assesses surveillance systems as a form of security that manages and polices national borders. Composed of artists Sharlene Bamboat and Alexis Mitchell, Bambitchell originally created *Border Sounds* for Nuit Blanche Toronto in 2011, and the artwork was later included in the group show *Border Cultures:*

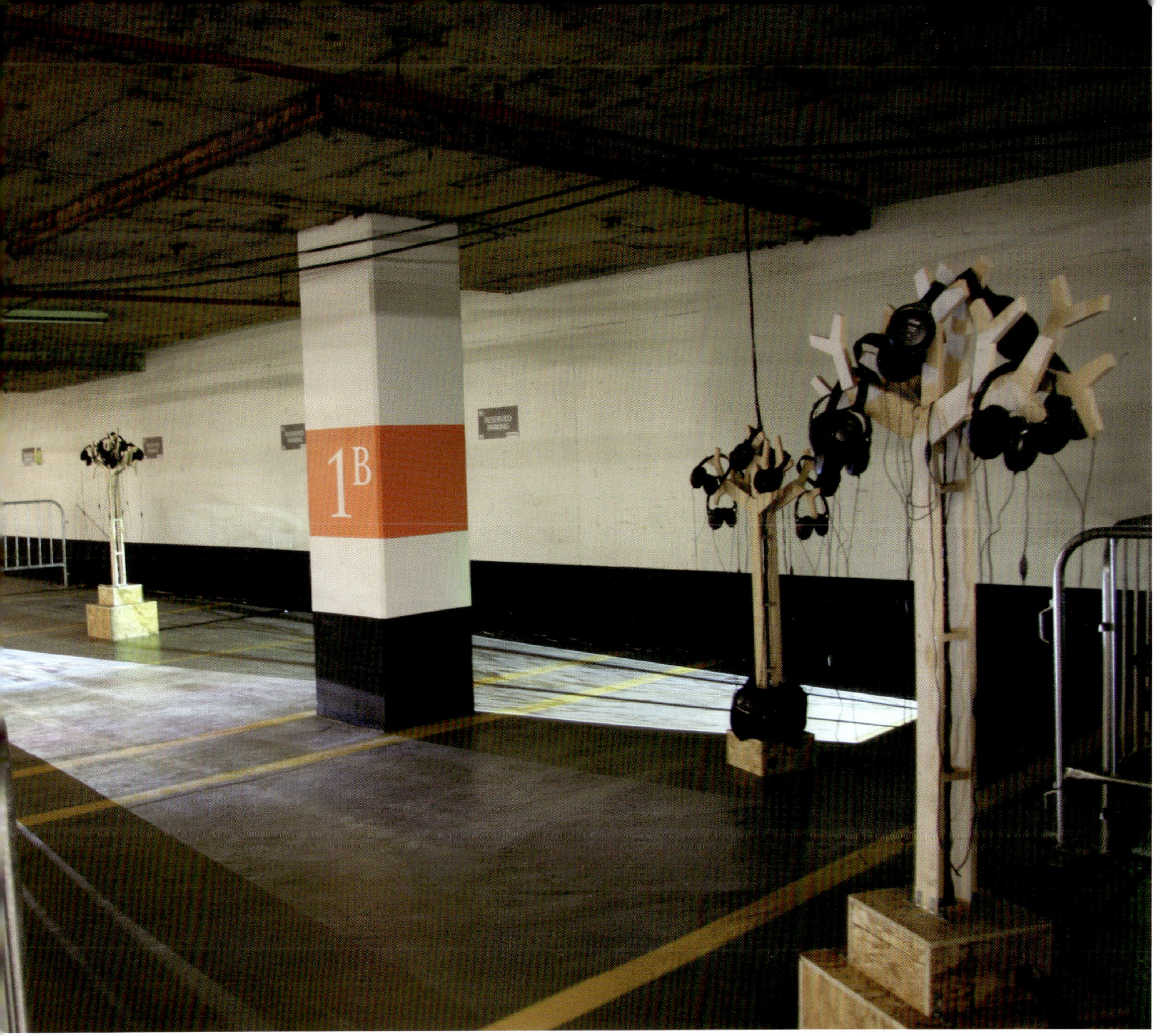

4.2 • Bambitchell, *Border Sounds*, 2011–15.

Part Three (security, surveillance) at the Art Gallery of Windsor in Windsor, Ontario in 2015. For Nuit Blanche, the installation was set up as an immersive experience in a cordoned-off section of a parking garage, with clear gateways for entry and exit to manage access and the movement of bodies. Five wooden structures resembling leafless trees with reaching boughs stand throughout the space, each with a collection of headsets mounted to their branches. Each structure represents a different state – the United States, Pakistan, Palestine, Israel, and Canada – and the correlating headsets play a voice reading the text of each station's passport or mobility documents in a flat tone, over a dubstep track. The artists chose dubstep because of its connection to the movement of people from Jamaica to the United Kingdom in the 1960s and 1970s, so the background music reflects the history of border crossing and migrations.[22] Bambitchell designed this piece as a video, performance, and sound installation in the form of a silent disco. Lights and video projections illuminate the walls of the space, adding a visual noise element to the art piece. Audience members move through the space of the installation and select stations to stop at and try on the headsets.

The artists selected the states represented in the piece based on their own personal backgrounds and histories. The five states they included demonstrate very different levels of internal and external access and mobility, which the artwork represents through the variable comfort and ease of engaging with the listening experience of each state. The cables of the headphones attached to each of the wooden structures vary in length, which determines how far and how much listeners can move when listening to the tracks. For instance, Canada's cords are quite long, which allows a relatively high level of movement, whereas those for Palestine are very short and force listeners to stand in close proximity to the listening unit and to each other when engaging with the music. The piece foregrounds Palestine as the only state without its own passport and highlights the inequities of transnational mobilities through the physical limitations of the audience's engagement with the work. This strategy of differentiating the mobility permitted in different territories makes material what Simone Browne articulates as "enactments of surveillance [that] reify boundaries and borders, and weigh down some bodies more than others, where the outcome is discriminatory treatment."[23] The artwork embodies these differences – via the different lengths of headphone cords – as a vehicle for audiences to, in Bambitchell's words, "encounter borders in visceral and affectively engaged ways that go beyond the intellectual treatment to explore what it means, and how it feels, to be moved and/or to be stuck in place."[24] The installation's organization coordinates a visceral and affective experience of engaging with the constraining mechanisms of different territories, raising questions of access and mobility.

According to the artists, they intend *Border Sounds* to challenge the conventions of border politics and policies by bringing forward "queer subjectivities [to] disrupt straightforward or chrono-normative narratives of 'belonging' and 'home.'"[25] Here, "queer subjectivities" function to create other modes of being and engaging with national frameworks that are not linear or normative. The disruption of "straightforward" or "chrono-normative narratives" occurs in the act of engaging with the artwork: it juxtaposes the joy of listening to upbeat music with the reality of the lyrics; the comfort in the act of listening on longer-length headphones with the restriction and discomfort of engaging with the shorter-length cables; and the freedom of listening to the music individually on separate headphones while being in proximity to strangers doing the same. To participate and engage in the artwork is to physically navigate the disruption of a comfortable and straightforward listening experience. However, creating a community of listeners engaging and interpreting together is key to the piece. It is the participatory togetherness, and even collective joy, that produces a sense of queer subjectivity that challenges the alienating experience of surveillant policing at national borders. As Bambitchell states, the purpose of *Border Sounds* is to align "with queer critiques of mobility because it takes into account the deeply subjective aspects of movement in relation to borders and border-crossings while foregrounding a sense of intimacy within processes commonly understood and depicted as only ever alienating."[26]

In the "sense of intimacy" Bambitchell creates within the piece, *Border Sounds* rejects the individuation and alienation of the surveillant policing state that determines unequal forms of inclusion and access. Bambitchell asserts queerness as a state of otherwise – that is, as a way to operate outside the normative expectations of assimilation and inclusion into national spaces. The piece works in conversation with gender theorist Jasbir Puar's conception of "homonationalism," which examines the project of US national and imperial expansion into queer life, particularly during the period of the War on Terror. For Puar, this amplification of forms of nationalism to include previously excluded subjects, such as 2SLGBTQIA+ people, generates a "brand of homosexuality [that] operates as a regulatory script not only of normative gayness, queerness, or homosexuality, but also of the racial and national norms that reinforce these sexual subjects."[27] *Border Sounds* rejects this "regulatory script" and seamless integration and acceptance into national narratives. The assertion of queerness in the art installation as both a positionality and political position that resists easy assimilation into state narratives is an explicit rejection of settler colonial, racist, and cisheteropatriarchal systems of the nation-state and the surveillant policing operations that support them.

Because *Border Sounds* juxtaposes the potentially dehumanizing process of border crossing with the pleasures and intimacies of music and dance, it both affirms different affective experiences of the surveillant policing of border spaces and creates an intimacy between audience members in thinking and feeling through these experiences as a collective endeavour. This work has a playful tone to it, but its meaning and purpose present a clear and important message about the inequities of movement and mobility among passport holders (and non-holders). These frictions are intentional. Just as the work critiques border policy while encouraging connection in public spaces, it similarly blurs the boundaries of inclusion and exclusion, access and denial, and connection and disconnection as it brings the intimacies and attachments of being in relation with one another to arrangements of border transit that are conventionally alienating by design. The piece derives meaning not only in the arrival and in-placeness of listening to the music and spoken text at each station, but also in the mobility of the in-between – of moving from station to station, place to place – and in the ease or unease this transit brings. The installation manifests how the documents and bureaucracy of surveillant policing – such as passports and other papers – grant access and mobility through and between different states and shows that intimacy, relationships, and being in place together can reframe and challenge the alienating space of and attachments to the surveilled border.

The third work in this section, Abdi Osman's *Passport* (2017, fig. 4.3), also uses embodied engagement along with the concept and aesthetic of passports as part of its critique of surveillant policing systems. *Passport* was included in a solo exhibition of Osman's work entitled *Variations of Black, Queer, and Otherwise: Works by Abdi Osman*, curated by Dina Georgis and Sara Matthews at the University of Toronto Art Centre, from 5 June to 17 July 2019. *Passport* is composed of four self-portraits photographically printed on large-scale linen sheets. One shows the artist's passport photo and cover, while the others depict him smiling into the camera, holding the passport beside his face. In each, the opened passport reveals the collection of stamps that mark his entrance and exit from various territories. This series builds on larger questions from the artist's broader creative practice, which Osman describes as addressing "black masculinity as it intersects with Muslim and queer identities."[28] The textiles hang in a line from the ceiling and are long enough to almost reach the floor. Since gallery visitors must walk in and around the hangings to fully apprehend them, the artwork engages them in an embodied encounter.

The size of the fabric is an essential element to the meaning produced in the piece. In her review of the exhibition, Black studies scholar Christina Sharpe writes that "these photographs are printed on large pieces of fabric that are about the dimensions of a flag; those flags that appear in airports and that announce nation – announce belonging and unbelonging; borders and surveillance."[29] According to interdisciplinary arts scholar Ellyn Walker, who wrote the curatorial essay for the exhibition, the hanging fabrics evoke "memories of checkpoints and border crossing, similar to what Osman might endure crossing a border as someone who is a Canadian citizen and who immigrated here as a Somali refugee many years ago. The immersive installation underscores the performativity of the passport to both engender as well as inhibit one's transnational movement and sense of national belonging."[30] For both Sharpe and Walker, the size and materiality of the medium brings together content

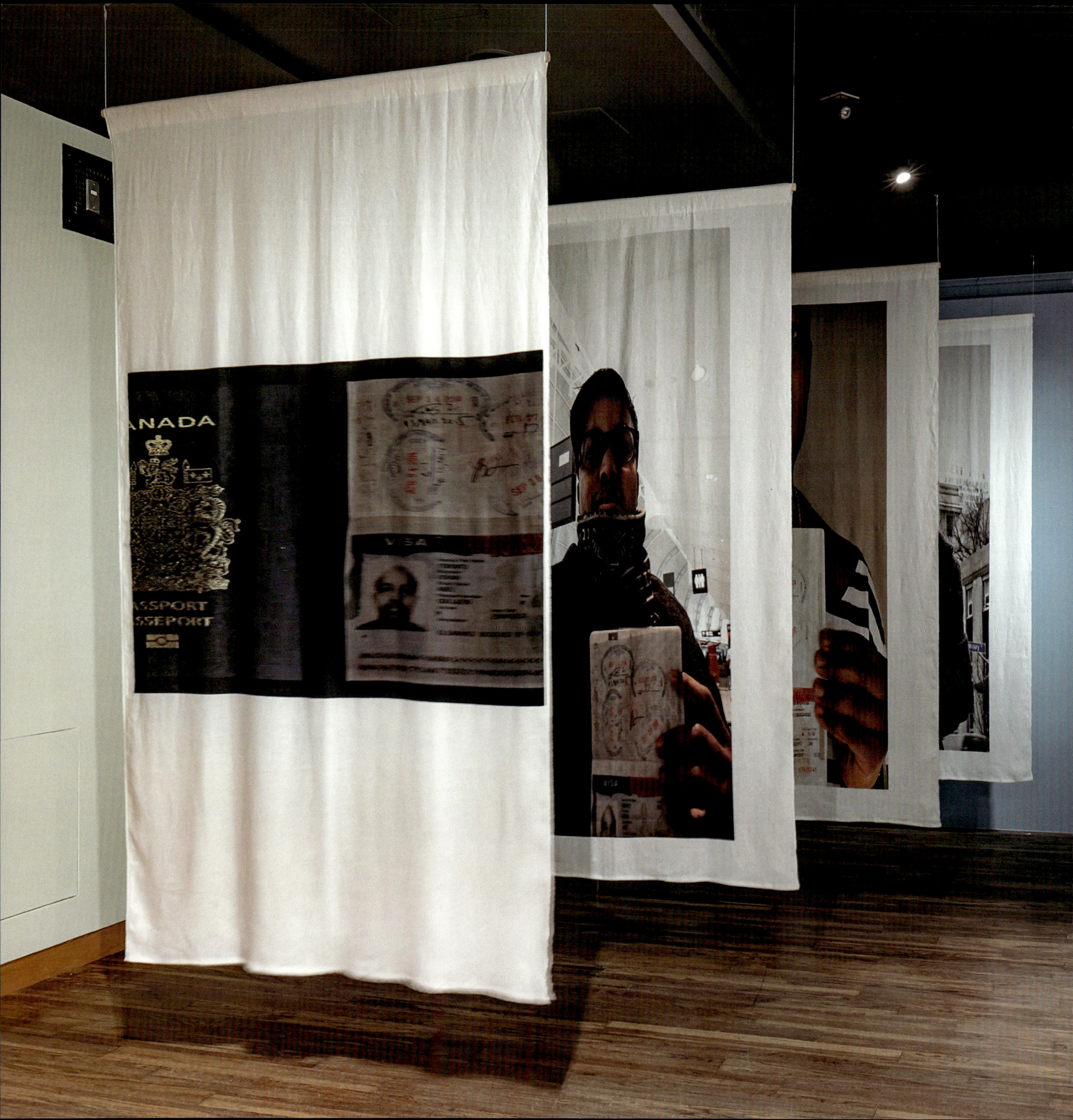
ANADA
ASSPORT
ASSEPORT
VISA

4.3 • Abdi Osman, *Passport*, 2017.

and form in a gesture that poses a keen critique of how and for whom the passport operates as a surveillance device for national belonging.

In the space of the exhibition, the movement of the viewer's body causes movement in the fabric, resulting in a dynamic, kinetic relationship between audience and image that evokes questions of intimacy, proximity, and relationality. As audience members walk through and among the fabric photographs, their movements as they engage with the material and visual aspects of the work mimic Osman's as he would have navigated checkpoints and borders, the passport as his official ticket that approves, monitors, surveils, and enables access. The passport *claims* Osman, declaring his belonging and attachments to Canada. Yet, notably, Osman's passport also disavows or withdraws belonging as well: as Walker highlights, Osman's passport reveals a ten-year visa "that enables but does not guarantee entry into the USA. It speaks to the heightened state of surveillance, particularly as it relates to Black and Muslim bodies travelling from one place to the other."[31] In *Passport*, Osman takes up a resistant and critical vision to interrogate systems of surveillance and access that yield inclusion in or exclusion from the settler colonial nation; in this, he questions the ways in which these systems and their agents inequitably target certain people and communities.

The passports themselves function as a surveillance technology, particularly when it comes to the movements of Black, Muslim, and queer people. As Sharpe writes, "surveillance technologies are always changing but the changing same is the degree to which Black people are surveilled – the ways that additional papers or additional screening are almost always required. The 'joke not-joke' about the ways that if we aren't immediately sent to secondary screening because of a name, where we've travelled or lived, national ancestry, etc., then we are almost always 'randomly selected' when we travel. Hands and bags swiped for explosives, put through the backscatter X-ray machine, etc. Secondarily screened. Stopped."[32] Sharpe outlines the ways in which passports are treated surveillantly as stand-ins for bodies, specifically for Black bodies; if the state and its arms are not scanning the bodies of Black people directly, then they are scanning their passports for names, origins, or clues about their bodies. Here, Sharpe critiques the biases of surveillant policing for its inequitable deployment based on white supremacist assumptions of whose bodies require additional monitoring to ensure national safety and security. In his artwork, Osman takes the restrictive impulse behind these racist, anti-Black, Islamophobic, and homophobic surveillance assumptions and turns it on its head by revealing in large-scale public format his access and permits to move, belong, and be in place. Osman turns the surveillance structure against itself, revealing rather than concealing. He reclaims the passport as evidence of his own experience rather than as an external marker that validates his presence and belonging. In this way, with *Passport*, Osman personalizes the surveillant policing system, which seeks to categorize, stereotype, and dehumanize Black, Muslim, and queer people, by using its own papers to claim his personal history and narrative of belonging.

The three works in this section all speak to surveillance operations that govern access, specifically the passport as an agent and tool that grants or denies both movement and belonging. Despite claims of widespread application in the name of national security, these operations are applied unevenly, based on socio-political assessments about which people and communities both external and internal to the nation are presumed to pose a threat. As Browne asserts, "these technologies of border control (passports, biometrics, airport preboarding passenger screening zones) are developed within, put to use, and often replicate existing inequalities."[33] The creative works here illuminate how these protective surveillance forms operate unevenly for different people, groups, and communities, and they question issues of access and belonging, particularly to the Canadian settler colonial nation-state. In *Papers Please!*, the artwork creates an encounter for broad publics to notice how they are implicated in surveillant policing systems that outline exclusions; *Border Sounds* creates spaces of togetherness that aim to challenge the alienating space of the surveilled border; and *Passport* reclaims the passport as a record of one's own personal history, rather than purely a surveillant policing device that can deny inclusion. Taken together, these works shine a light onto the problematic context from which these surveillant policing measures are produced and applied, and reveal the underlying racist, cisheteropatriarchal, and white supremacist assumptions that dictate where and how these measures are taken. In finding, variously, common discomfort, common joy, or common affirmation of belonging, the artworks bring audiences together to conceive of alternative pathways toward reclaiming oneself and belonging to communities outside the categories demanded by the settler colonial state.

Intimacy

The artworks in this section build on concepts of belonging to examine the idea of intimacy as a form of knowing and being together that is antithetical to surveillance logic. Whereas state and corporate surveillance in the service of settler colonialism emphasizes the individual above all else, and thus functions to seed competition, distrust, and suspicion, intimacy serves as the relational and affective bond that brings people together in trust and care. Intimacy as an affective, social, and political concept has been taken up by a number of activist movements, including those fighting against white supremacy, racism, ableism, and cisheteropatriarchy.[34] The works in this section share an affinity with some of these activities and relevant scholarship in that they take up intimacy as a form of knowing that involves closeness and emotional proximities and reveals vulnerabilities to be seen, heard, and cared for as a challenge to systems of violence and oppression. In contrast to underlying belief systems of settler colonialism, which dehumanize and isolate people from community in the service of domination and control, intimacy as a political concept is centred on a politics of inviting in, of trusting another enough to reveal yourself, thereby affirming relationality, interdependence, and collective care as modes of resistance. As cultural theorist Lauren Berlant writes, intimacy exists as "a narrative about something shared, a story about both oneself and others that will turn out in a particular way. Usually, this story is set within zones of familiarity and comfort."[35]

The two artworks in this section play with intimacy in relation to these "zones of familiarity and comfort" by situating surveillance in relation to the space of the home. The first work, Steve Giasson's installation *VOX* (2015–16, fig. 4.4), uses the space of the home as a barrier to knowledge – as a way to critically re-enact the fragmentary and limited access of surveillance to our intimate lives. The second, Derek Dunlop's photographs *A Lover's Geography* (2008, fig. 4.5), uses the home as a generative space of intimacy and insider knowledge. These two artworks position intimacy as an affective space of knowing in opposition to surveillance logic, which seeks to extract and hoard personal information as part of a settler colonial system that demands access to private knowledge without the promise of trust and care. Surveillance here is a distanced and distancing form of knowing, a position that limits any depth of understanding of the nuances of the details, the interpersonal, the emotional, produced through the knowledge gathered. Seen in this way, the space of intimacy highlights the limits of the information gathered through surveillance capture, emphasizing areas central to one's identity that cannot be obtained and distilled from a distance. These artworks, then, both resist surveillance logic as a vehicle for knowing and envision spaces that move beyond surveillance by emphasizing intimacies as a space outside its gaze.

Giasson's *VOX* creatively reframes surveillance logic by creating narratives from broken fragments and highlighting the gaps and limits to surveillance systems. The installation involves two hundred pages of transcribed text Giasson recorded from conversations he overheard outside his apartment over the course of a year, from 25 July 2009 to 24 July 2010. The work was featured in the Palais des congrès in Montreal for the Festival Art Souterrain in 2015 under the theme "Security in Our Society: What Remains of Our Personal Freedoms?" The work was exhibited again in Montreal at the 2-22 Building from 9 August to 9 October 2016. With *VOX*, Giasson translates the overheard conversations into visual form by printing the text on transparent sheets mounted over large windows in the gallery space. The design of the piece is meant to suggest the shape of a soundwave: the text is centred in columns on the sheets, allowing the unjustified margins on both sides of each column

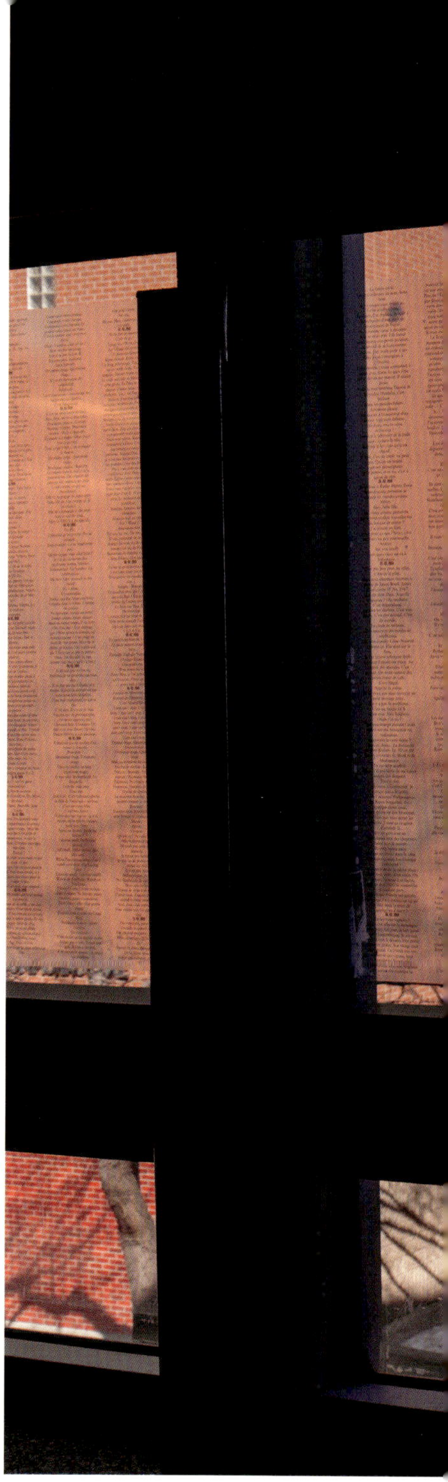

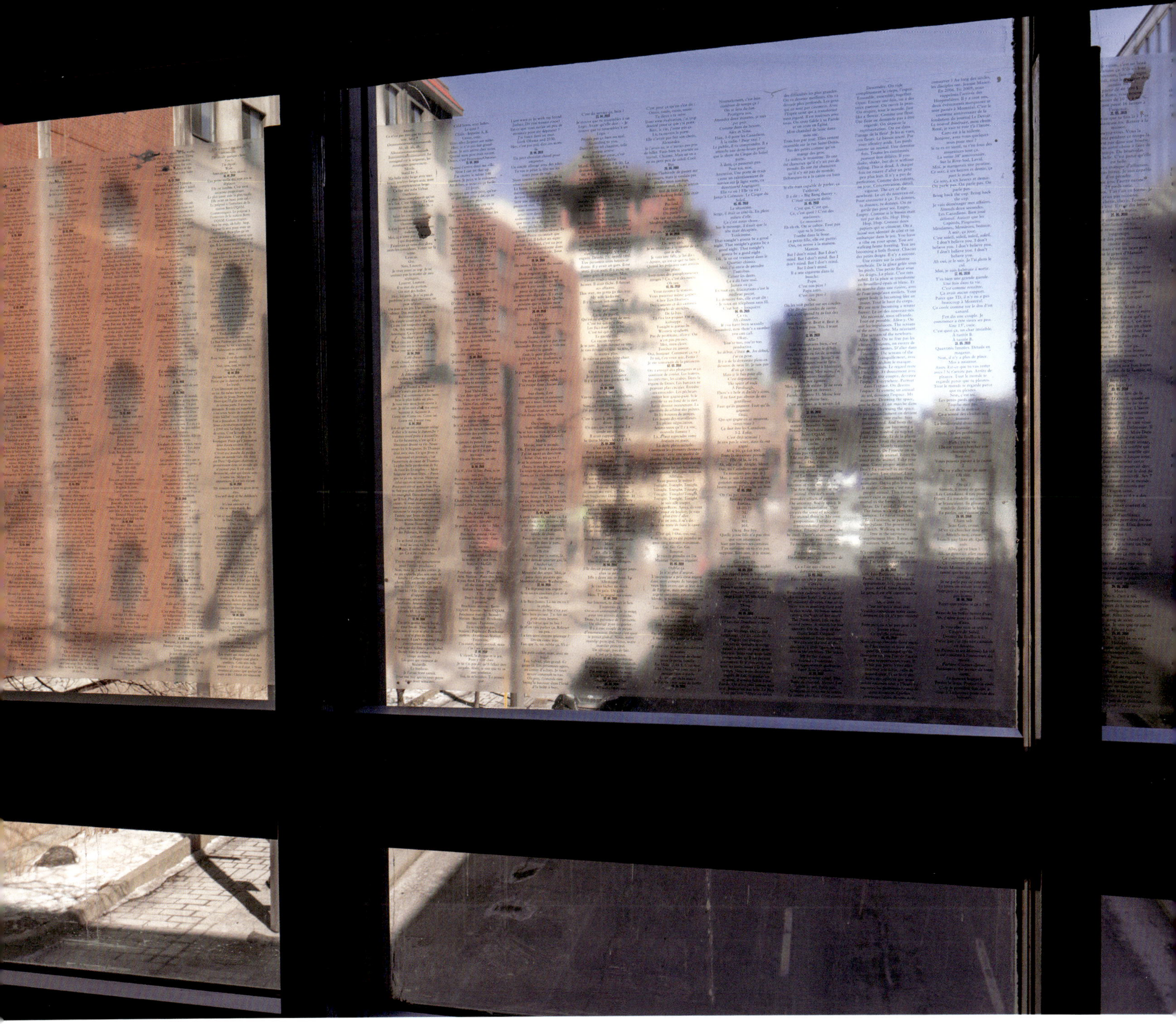

4.4 • Steve Giasson, *VOX*, 2015–16.

to undulate and curve. This form gives visual movement to the spoken word, but it does not replicate the affect or conversational gestures in the moment. Rather, the form of the installation produces many possible meanings in its relation to the written transcripts of overhead conversations. The transparent sheets create levels to the viewing experience because the visitor must decipher the text against the background they see through the window. In this way, Giasson creates multilayered frames through which the audience must interpret meaning: the overheard conversations; the recorded words as text; the viewing in the gallery space; and the looking through the window to the world outside the display. This device visually brings the external world into the display itself, in much the same way surveillance capture brings extrinsic information into the surveillance context.

The transcript of overheard conversations between unwitting participants-as-performers created by a man sitting in his house by a window overlooking a public street represents the anywhere-everywhere of surveillance operations. I see the activity of *VOX*, in many ways, as an analog replication of advanced surveillance technologies such as NextDoor (a social app for neighbours to report and discuss local events), Neighbors by Ring (an app that is an extension of Amazon's Ring and offers real-time crime reporting and alerts from law enforcement), and Citizen (originally called Vigilante, an app that provides real-time safety and crime alerts). These mobile and online apps have millions of users who report suspicious activity and surveil local areas for perceived troubles. Yet studies have shown that these forms of peer-to-peer and neighbourhood surveillance are "producing reports riddled with dog whistles and overt racism" and essentially "erode our sense of safety and social trust, substituting racial profiling, vigilantism, and neighbor-on-neighbor spying."[36] *VOX* reminds us that surveillance use is not reserved purely for Big Tech or corporate or state actors; the work's lo-fi surveillance instruments reflect the internalization of Foucault's idea of disciplinary power, where power is no longer strictly top-down as a matter of domination and submission but, rather, is diffused through the population so that the people and the publics function to monitor and contain one another.[37] Rather than a salve or utopian vision of looking back at structures of power and control, these forms of peer-to-peer and neighbour surveillance can actually replicate the biases and prejudices found widely within a settler colonial society like Canada.[38]

At the same time that the work mimics the ubiquity of surveillance, it also reveals that surveillance can never capture everything. Giasson highlights what aspects of a surveilled moment get lost within its capture: the flat form of the spoken-to-written words offers no context, tone, colour, or character that comes with lived experience. This lack of nuance illuminates the spaces in between that cannot be easily monitored, captured, and conveyed as part of the surveillance apparatus. In this way, the piece asks what is lost in translation from lived experience to the surveillance record, highlighting how both the record and the knowledge it produces are reduced to a two-dimensional form. Significantly, this piece also conveys the intricacies of interpersonal and social relations – the patterns of identities, relationships, and intimacies that cannot be grasped through surveillance capture. While Giasson can offer to the art audience the details of overheard conversation, he cannot fill out these fragments with any true knowledge or insight into the contexts, feelings, or background of the people or the topics. *VOX* creates an archive not of conversations overheard but of Giasson overhearing them. It emphasizes the artist; and, by extension, it emphasizes the art audience, who comes to the piece through the artist's eyes as surveillance outsiders. Rather than surveillance offering insights into the space and private life of another, surveillance here operates as a barrier, a limit to accessing and being invited into the intimate life of others. And while the artist and we as the audience may try to fill out the gaps and fragments of overheard discussions with our own perspectives, guesses, biases, and narratives, we are essentially locked out of the interpersonal intimacies that generated these conversations.

Giasson invites the art audience inside his own home, but only to join him in looking out. In many ways, *VOX* highlights the antirelational logic of surveillance, one that operates outside of belonging, togetherness, and community. As part of a settler colonial system that prioritizes whiteness, ableism, and cisheteronormativity as a hegemonic identity, the surveillance gaze fundamentally lacks intimacy and care. While the artwork generates a relationship between artist and audience, it is one of loneliness, limits, and being in relation in solitude. This work is a portrait of contemporary life in a historical moment that places challenges on social conceptions and expectations of private lives and intimate access. It reveals the limits of surveillance as an operation that gathers details but cannot access the spaces of intimacy required to build relationships and develop knowledge beyond those details.

Dunlop's photographic series *A Lover's Geography* works in conversation with Giasson's *VOX* to examine intimate spaces and surveillance knowledges. Unlike *Vox*, however, which emphasizes the limits of surveillance structures in extracting intimacies within social relations, *A Lover's Geography* creates a sense of intimate and insider knowledge – of being invited into a private, trusted space together. Included as part of the group exhibition *Cyber-Surveillance in Everyday Life* at the media arts centre InterAccess in Toronto, Dunlop's work creates a tension between the impersonal quality of the surveillance gaze, on one hand, and an intimate feeling of being in close proximity to another's life, on the other. *A Lover's Geography* is a series of pixelated photographs, largely of disheveled beds and couches, devoid of any physical human presence. Dunlop presents a narrative of a lived life by presenting zoomed-in features of a living space, often including a section of a chair, pillow, or bed somewhere in the frame. The spaces are slightly messy and uncurated, as if to suggest the intimacy of a home without the staged qualities that come with performing for outside eyes. The quality and composition of the photos – uncentred, unfocused, indistinct – lend the images an aesthetic located somewhere between crime scene photographs, family pictures, and webcam surveillance. Taken together, the photographs give the sense of a life surveilled, but whether that is done with love or malice is not immediately clear. To the extent that they reveal portions of a private interior, the images invite the viewer in, but, since the picture they represent is incomplete – literally and figuratively – they deny the viewer any real knowledge of the person or story attached to the images.

While no bodies are present, the images depict spaces meant for bodies to be – not only because the broader context is an interior living space but also because the frame focuses in on materials designed specifically for holding or supporting

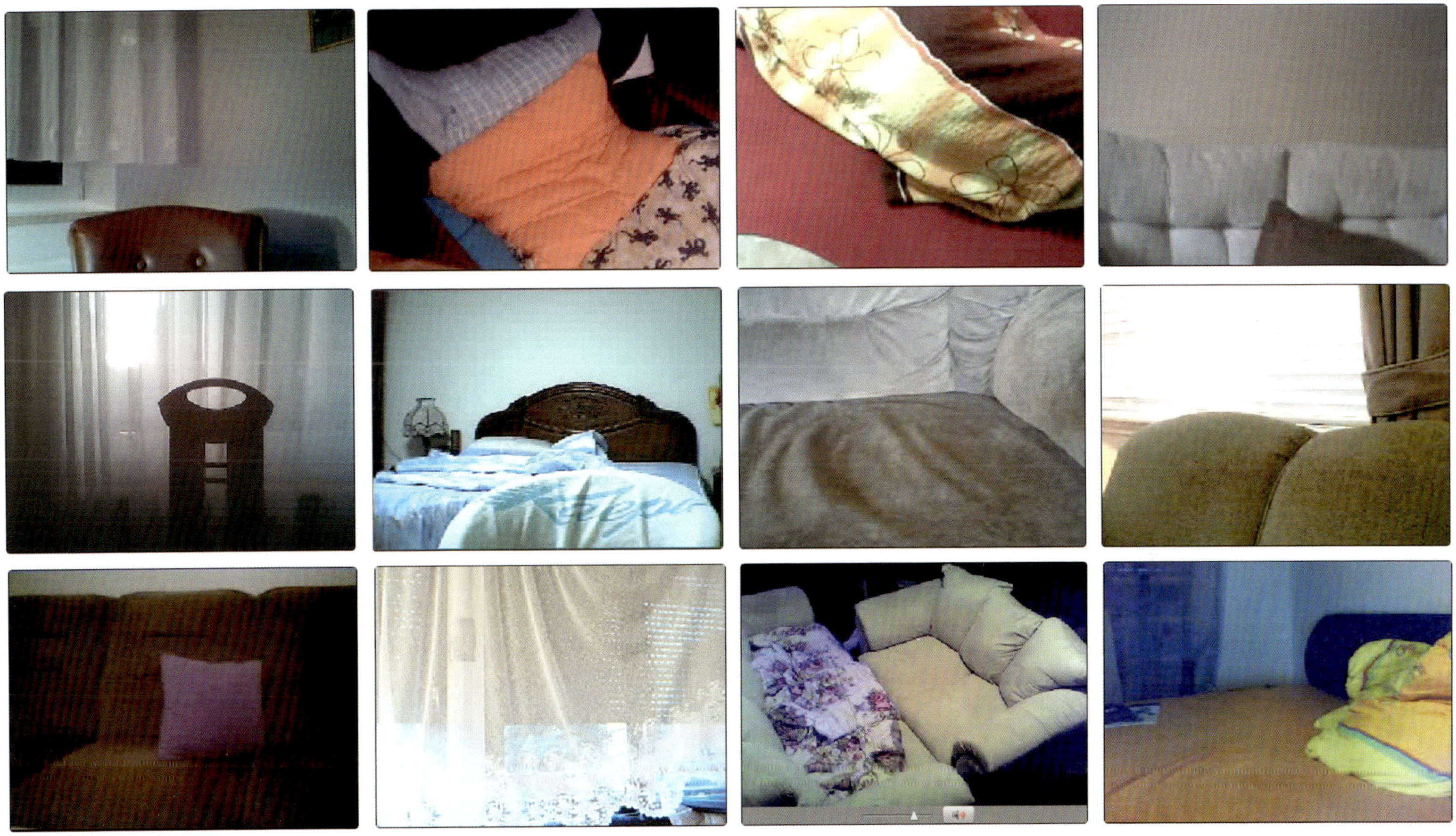

4.5 • Derek Dunlop, *A Lover's Geography*, 2008.

bodies: beds, chairs and sofas, pillows and cushions, and blankets. These items become stand-ins for the sensibility of bodies existing within intimate spaces, in private, in comfort. In this sense, *A Lover's Geography* opens up aspects of the private space of the home, where the body is relaxed and at rest. The surveillance angle of vision taken from the perspective of a webcam conflates intimate yet incomplete portraits of the home with the all-seeing, prying eye. The title of the piece, *A Lover's Geography*, suggests that the monitoring eye is that of a lover – one who is privileged with access and insight into the intimate geographies of a private life. The work resists and reconceives surveillance by imbuing it with a sense of ambivalence: it reappropriates the aesthetic of surveillance associated with state and corporate extraction of personal information and data from a distance, and blends it with indications of interpersonal surveillance that attends to a loving and cozy knowledge of life and the home.

This approach to a surveillance point of view more broadly attests to what Hannah Godfrey describes as Dunlop's artistic interest in a queer way of looking, one attached to forms of cruising. Godfrey explains that "the act of looking that is required by cruising situates the person who is doing the looking as a body open to and seeking particular experiences. The gaze has intention: it is evaluative and holds the desire for exchange and gratification; it carries expectation, even that of simple rejection."[39] With this expectant gaze in mind, Dunlop's work takes on another layer of meaning as he re-presents the surveillance gaze as one of queer interest and desire. *A Lover's Geography* subverts the expectations of the dominant surveillance gaze, one associated with settler colonialism and its cisheteropatriarchal logics that have long policed, monitored, and enacted violence on queer people and the spaces they carve out for themselves. This work can be seen both as a loving testament to insider knowledge when invited into the intimate and private life of a space or person and as a deliberate refutation of outside viewers who are not fully invited into the scene. This refutation is itself a form of resistance against the surveillance apparatus that demands full disclosure of private moments as a supposed safeguard to the Canadian nation-state and its citizens. This artwork illuminates the spaces of the unknown of surveillance capture, while optimistically reappropriating surveillance codes toward another, more subversive purpose.

By depriving the viewer of information about who the home, or homes, belongs to, this artwork examines how narratives and understandings are produced through looking with limited context. What is rendered absent? Whose life – or lives – is represented? What does it mean to look without knowing? Or to recognize oneself as outside the central narrative? The spaces of unknowing, of absence, of non-understanding, are as present as what can be known, or seen, or understood. *A Lover's Geography* reveals the gaps of knowledge of surveillance structures – the ways in which understandings or narratives are constructed through fragments or pieces. In many ways, the message of the piece is about what cannot be represented, and how viewers fill in empty space with their own narratives. This artwork highlights the spaces in between, where the detail gathering of surveillance falls short and the possibility of intimacy reveals its strength.

Both *VOX* and *A Lover's Geography* play with exposing the interior workings of people's lives for the surveillance gaze of the art audience. They take the familiar aesthetic of the surveillance gaze – the penetrating lens of knowledge intended to gather and extract information – and they reframe it. The spaces the artworks generate show both the limits of surveillance and the possibilities of being in relation and beyond the reach of the surveillance gaze. While the artworks share these themes, they approach intimacy and the home in different ways: whereas *VOX* positions home as a place where safety and intimacy are not necessarily guaranteed, *A Lover's Geography* conveys home as providing an oasis in which to build forms of care and togetherness outside the surveillance gaze. Both works ask, can you have intimate knowledge without intimacy? How can intimacy be utilized as a strategy to resist and rethink surveillance structures in their service of settler colonialism in its intersection with cisheteropatriachy? Indeed, these works resist such systems and their surveillance operations that situate knowledge as singular, individual, and disconnected: Giasson's *VOX* emphasizes surveillance as a limited form of access because of its lack of intimacy, while Dunlop's *A Lover's Geography* builds the possibility beyond surveillance vision for a queer politics to enable different forms of being together. In these reimaginings, both works highlight what writer and theatre director John Edward McGrath identifies as the power of creative practice in relation to surveillance: that artworks can "explore the experience of surveillance – how we feel and live within what we might call surveillance space."[40] Taken together, they work to provide sites that counter mainstream narratives embedded in the contemporary surveillance space, and, especially in the case of *A Lover's Geography*, they imagine possibilities of finding relations and intimacies to actively create other ways of being in the world.

Community

The act of creating community can be viewed as an act of resistance in itself, as community sits in opposition to settler colonial surveillance apparatuses that prioritize the individual and disrupt connections between people and their communities. The transformative potential of art, in this context, can lie in its capacity to actively involve communities in the creation and interpretation of artistic expressions. The four creative projects in this section work in conversation with strategies of intimacies explored in the preceding section to develop communities of care as a strategy of resistance to surveillance. By embracing the principles of co-creation, collaboration, and dialogue, these artworks strive to empower individuals and communities, amplify marginalized voices, and resist top-down corporate and state surveillance structures.

These works activate what art historian Claire Bishop describes as the potential for creative practice "to restore and realize a communal collective space of shared social engagement."[41] The art projects in this section use strategies to critique and look back at systems of power and oppression, but also, significantly, to mobilize art audiences as community members to imagine beyond surveillance structures that uphold Canada's project of settler colonialism and its systems of oppression. As

Browne articulates in relation to the power of creative works of surveillance by Black artists, in "certain acts of cultural production we can find performances of freedom and suggestions of alternatives to ways of living under routinized surveillance."[42] The artworks in this section invite the audience in as actors, actively engaging them in creative and critical scenarios that reposition their understanding of and relation to surveillance. These works humanize complex issues, shedding light on marginalized and silenced voices and addressing systemic injustices. The goal of these artworks is to pull back the curtain on particular operations of surveillance to reveal the unjust and violent logic of these applications, and then to build interconnections and create communities of care and engagement that inspire resistance and change.

Genevieve Robertson and Jay White's collaborative projects *Walk the Line* (2014–15, fig. 4.6) and *Watchers* (2015, fig. 4.7) draw on grassroots initiatives speaking out against a proposed oil pipeline that would cut through several communities in western Canada. By fostering collaboration between artists and communities, these projects nurture a sense of shared responsibility, belonging, and social cohesion. For *Walk the Line*, Robertson and White invited "artists, activists, neighbours and concerned citizens" to join them on a walk along a portion of the proposed Kinder Morgan pipeline, a route set to bring crude oil extracted from the Alberta tar sands to the west coast for transport internationally.[43] The purpose of this artwork, which the artists describe as "participatory walks," was to immerse community members in the land that might soon be altered by the impending project. In the words of the artists, "through our bodily presence, the intention was to internalize the spaces that the pipeline would intersect and affect; to engage in the ecological implications of the pipeline expansion; and collectively consider the complicated nature of this contested landscape."[44] This community-based art project contributes to other grassroots resistance efforts raising concerns about the pipeline at multiple points in its proposed route. Certainly, many Indigenous groups upon whose lands the pipeline would be built have undertaken a long-term and intensive resistance to the activities of the Texas-based Kinder Morgan company and the federal government that supports it. Despite widespread active resistance against the project, the

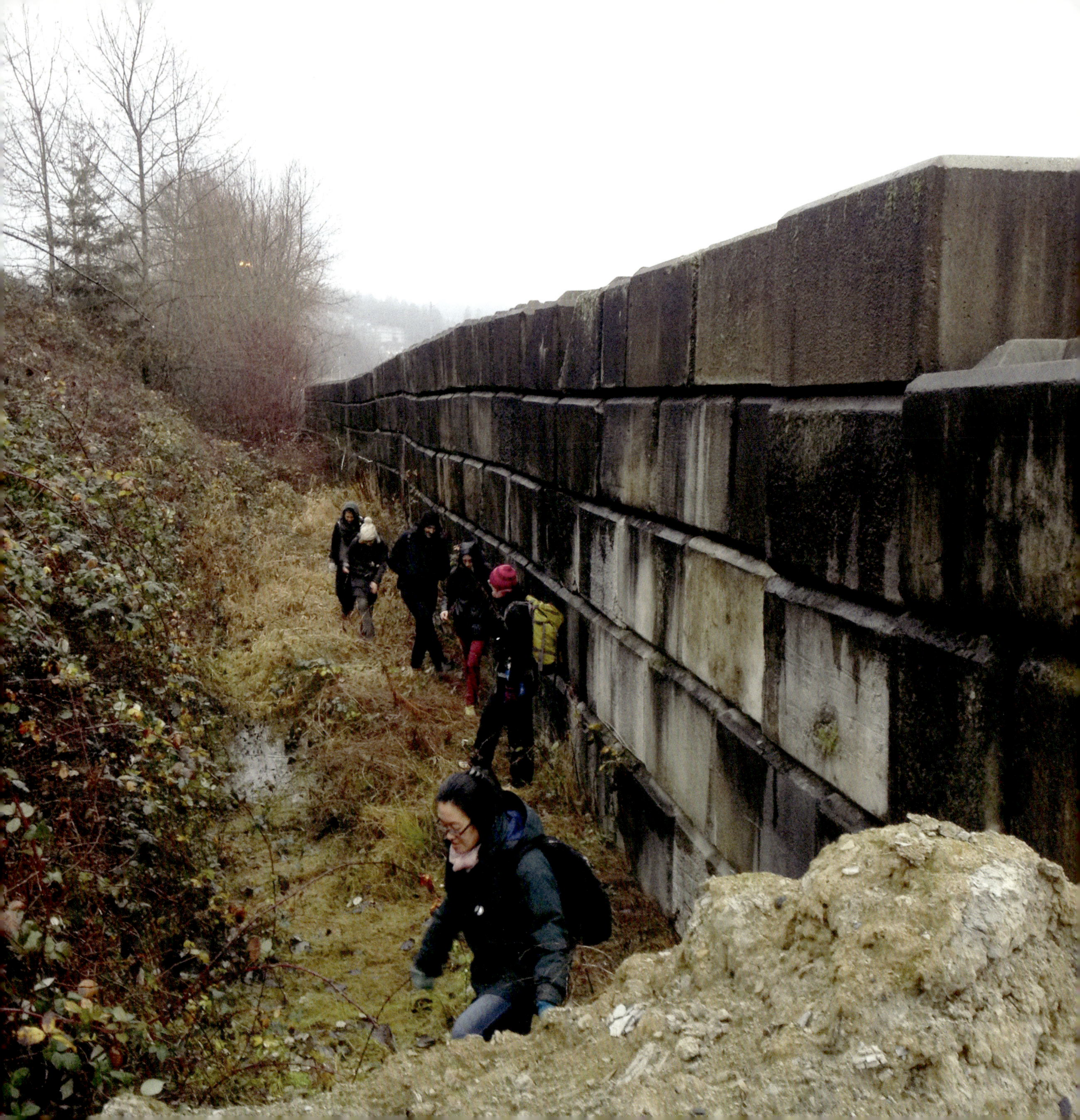

4.6 • Genevieve Robertson and Jay White, *Walk the Line*, 2014–15.

federal government purchased the pipeline in 2018 and renamed it the Trans Mountain pipeline.[45]

Watchers, built on the ideas underpinning *Walk the Line*, is a series of photographs that grew out of a second community walk, which took place on 20 May 2015.[46] The photographs bring the community participatory action of the *Walk the Line* project together with an oppositional confrontation with the surveillance apparatus that structures the policing and governance of the pipeline plan. To make the images, the artists mounted six-week exposure pinhole cameras along the proposed pipeline route as a response, according to the artists, "to industry-government collusion on the restriction of [citizens'] agency and ... [as] a gesture towards resistance, counter-surveillance, and seeing through extended temporal perspectives."[47] The mounted cameras trace the same route as the previous walking project but introduce the presence of technology, surveillance, and distanced monitoring as part of meaning making. The pinhole cameras plot the route from their perch on trees in order to record and capture the space and any activities on it. Having mounted cameras in place resists and circumvents police operations, as protestors were previously arrested for taking photos of the proposed pipeline route. This strategy to install the cameras in this way allowed the artists to produce images despite state and corporate measures to counter such activity.[48] Even though they generate somewhat crude, imprecise images, the pinhole cameras serve as a "counter-surveillance" gesture, recording the land as it is and providing a record of any changes to it from the proposed pipeline construction. This work turns the surveillance lens on corporate and state surveillance operations that are set up to police anyone who might challenge the project. It also provides visual access to the area to community members and activists who want to monitor any pipeline progress.

In keeping with pinhole camera photography, the resulting photographs are blurry and largely indecipherable (see *Watchers #3*, fig. 4.7). In these dreamy, unfocused images of landscapes, only certain obvious elements are recognizable: trees against a sky backdrop, a road snaking across the frame. The photographs of *Watchers*, therefore, represent more of a gesture to the land than a definitive or exacting mapping of it. This gesture connotes the murky politics of state and corporate expectations of the land-as-resource. The route of the pipeline is contested as crossing unceded Indigenous lands and forms the basis of a triangulation of claims between multiple Indigenous nations, the settler colonial state, and a private corporation. The photographic gesture of *Watchers* also attends to what can and cannot be captured – what is felt and experienced versus what is represented and quantified. The photographs recall the embodied experience of walking and being together on the land as a community in the *Walk the Line* project, which itself is meant to highlight the wider community connections of being in place, being in relation, and being together with the land in a way that rejects the extractive logic of settler colonialism. Surveillance images can reveal some aspects, but not the full picture, and *Watchers* depicts the inaccuracy of these technologies and the way in which they obscure what they record. So while the cameras and the resulting images grant community members access to seeing and tracking the land and the activities on it, the images also highlight the imperceptibility of knowing and understanding the land through the surveillance lens: they are a distant second to the experience of being on and with the land in community. In this way, *Watchers* photographs highlight the complexities of surveillance watching and the imprecision of capturing spaces and narratives, and, thus, they resist surveillance attempts to delineate and dominate the land with any illusions of confident clarity.

The complexities, in this instance, are certainly not abstract: based on surveillant policing actions, the Royal Canadian Mounted Police have arrested and charged a number of community members and activists who have protested and rejected the pipeline project.[49] Certainly, many Indigenous people – including the Squamish, Tsleil-Waututh, Coldwater First Nations, and a collective of bands within the Sto:lo Nation – have been at the forefront of these activist activities and have been targeted and arrested for their stance.[50] Real individuals and groups are feeling the realities of corporate interests that are protected by a militarized state apparatus. *Watchers* illuminates the blurry policies and procedures that form the basis of surveillant policing, which is organized and managed by both state and corporate forces and is

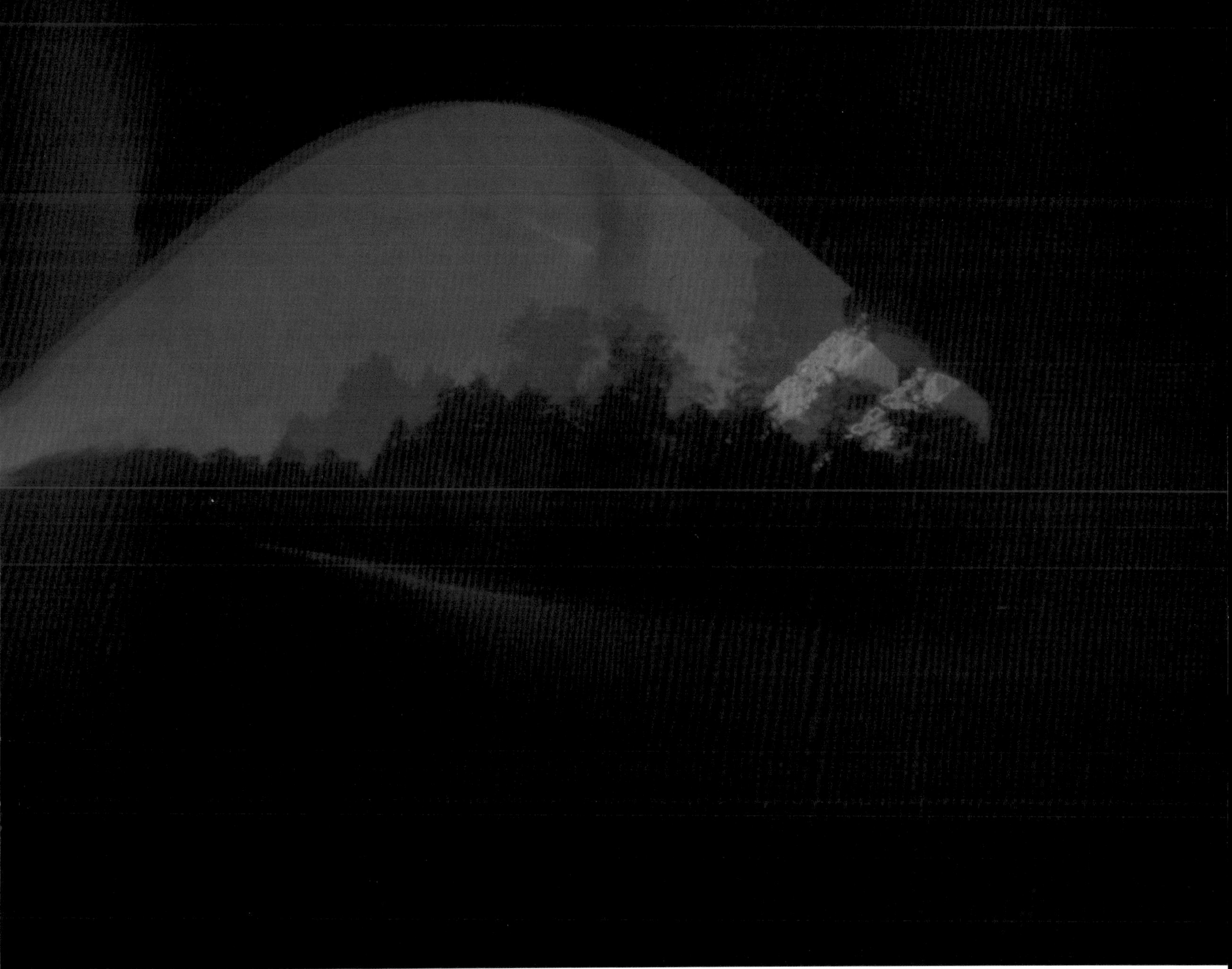

4.7 • Genevieve Robertson and Jay White, *Watchers: Watchers #3*, 2015.

designed to police and contain any community resistance to the crude oil project. The title of the piece, *Watchers*, draws attention to this surveillance while simultaneously flipping the presumption that the surveillant policing of this area is the purview of only state and corporate agencies. The work resists the idea that the area can be tracked and monitored only by these agencies – an idea that the police have enforced with arrests – and grants the community access to seeing and engaging with the space. In that it puts the power of looking and monitoring into the hands of the people and communities concerned about the land, *Watchers* illuminates the power of mirroring the lens of state and corporate surveillance to reveal what they do and do not show. It also reveals the potential of community solidarity and collective knowledge to resist the policing and narratives imposed by corporate and state surveillance, which aim to legitimize the claims to land and resources central to the settler colonial project.

Walk the Line and *Watchers* demonstrate knowledges of and relationships to the land that cannot be captured or quantified by the surveillance lens, and they reframe the possibilities of surveillance watching as an activist tool. The absence of bodies in *Watchers* emphasizes the narratives of the land that are so central to surveillance logic, bringing to light contemporary surveillance in the service of land exploitation and, by extension, the continuing legacy of settler colonialism and its extractions within Canada.[51] While these works resist surveillance logic through an emphasis on community, and being in place and on the land, other creative projects prioritize the body and social relationships in order to resist and reimagine community relations against and outside of settler colonial systems. Syrus Marcus Ware's drawings in *Their Borders Crossed Us* (2015) present detailed and intimate portraits of Black activists fighting for social justice. Included in the group exhibition *Border Cultures: Part Three (security, surveillance)* at the Art Gallery of Windsor in 2015, the drawings portray three activists: Nzinga Maxwell, a Costa Rican migrant seeking asylum in Canada; Taui Green, wearing an "I Can't Breathe" mask (fig. 4.8) to protest the murder of Eric Garner, an African American man killed by police using a chokehold; and Ware himself.[52] The catalogue for the exhibition describes Ware's work as developing "portraits of organizers/activists whose lives have been infiltrated by some form of

4.8 • Syrus Marcus Ware, *Their Borders Crossed Us: For Eric*, 2015.

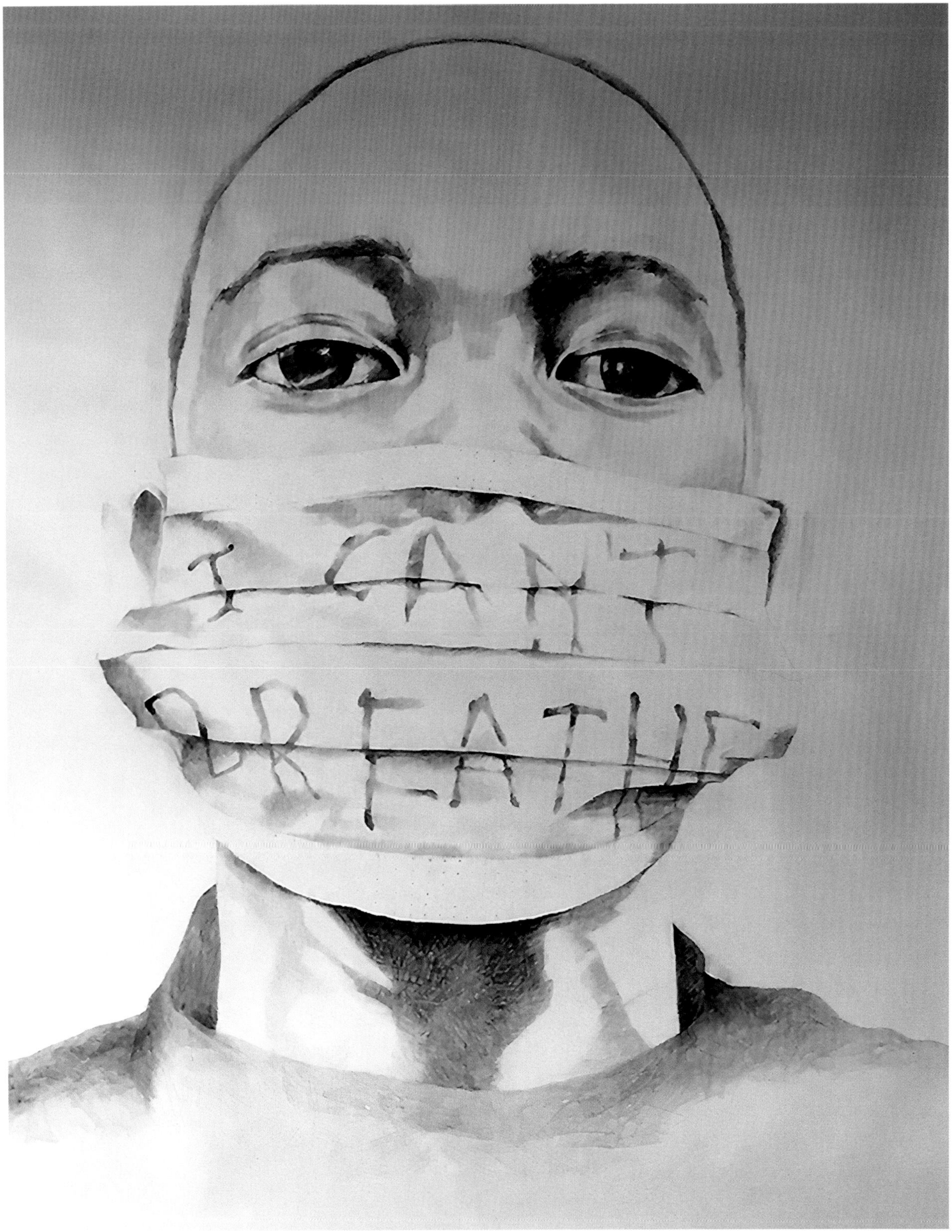
I CAN'T
BREATHE

surveillance measures in Canada."[53] The works in *Their Borders Crossed Us* are careful, intimate depictions in close-up portraiture completed in large scale. Showcasing three activists fighting for refugee, Black, queer, trans, and disabled communities, Ware's drawings provide a reframing of the surveillance lens that disproportionately and violently targets these communities and those who fight for them.

Their Borders Crossed Us shines its light squarely on those people who are inequitably scrutinized by the surveillance lens of policing and other state agencies that seek to silence and suppress particular forms of dissent. It flips dominant forms of surveillance looking by showcasing the same people who are often targeted and criminalized because of racism, colonialism, homophobia, transphobia, and ableism, and introducing intimacy and care as a strategy to emphasize them as heroic, admired, and cared for. The work also highlights the real-life dangers and violences these activists must navigate. For instance, someone reported Maxwell, one of the people depicted in this series, to Toronto Metropolitan University campus police, who, working with Citizenship and Immigration Canada, arrested and deported her.[54] The larger story to this portrait reveals the lived terror certain people face within settler colonial systems that determine which bodies belong and which do not.

According to Ware, these portraits can take up to eighty hours to produce.[55] This style of drawing is a laborious process defined by care, detail, and precision, in contrast to the widespread mass capture that characterizes many surveillance technologies. By concentrating on the slowness, time, and focus that builds a relationship between artist and subject, Ware's process challenges the mass gathering of data by broad-scale surveillance activities. It is in these details, as well as the resultant image, that *Their Borders Crossed Us* resists surveillance logic and its violence. This creative production highlights the spaces of community, care, and intimacy that challenge the dehumanizing and individuating surveillance structures in the service of systems of oppression. The portraits in *Their Borders Crossed Us* encourage viewers to see the person represented as someone elevated to a position of reverence through the art historical device of portraiture rather than as a figure reduced to a profile defined by criminality or suspicion through the surveillance image.

For Ware, thinking of his art as a form of activism and resistance is part of its inherent process: "My art is about my activism, my activism is an art."[56] His artistic activism challenges the frame through which Black, refugee, queer, trans, and disabled people are represented, and inspires community building and participatory engagement. As part of the *Border Cultures* exhibition, Ware did a performance entitled *Activist Love Letters* (2012–ongoing, fig. 4.9), which "encourages the participants and viewers to think about their role in sustaining a movement and supporting their communities."[57] This project encourages gallery goers to create love letters to organizers and activists they admire, creating connections between the visitors' own experiences and the larger goals of Ware's work. Connecting people cultivates networks of solidarity that directly challenge the dehumanizing logic of surveillance and the people who are the most targeted. By sharing knowledge, experiences, and strategies, communities assert their agency and resist being reduced to mere data points or targets of the surveillance and settler colonial apparatus. Ware's *Activist Love Letters* also offers connections that allow for the recognition and celebration of activist interventions, highlighting the tangible work done to protect land, resources, and human dignity. In this creative and caring action, the artwork highlights the potential of relational networks to reveal alternatives to extractive, controlling systems and demonstrates the power of collective action against corporate and state surveillance.

These two works – *Their Borders Crossed Us* and *Activist Love Letters* – together frame activism as a labour of love, care, and compassion, directly challenging the dominant narratives that portray activists as dissidents, criminals, or anti-social. Ware's artistic practice encourages connection, empathy, and kinship with communities of activists whose work and being are consistently overdetermined and undermined by surveillance. Ware actively brings the art audience into a larger community of those working toward change. He humanizes the subjects of his work and amplifies the collective bravery, resilience, and care embedded in activist practices. His artworks offer alternative narratives and imagery, countering the reductive, dehumanizing representations of state and corporate monitoring, and demonstrate that activism is as much about recognizing and sustaining community as it is about challenging systems of power.

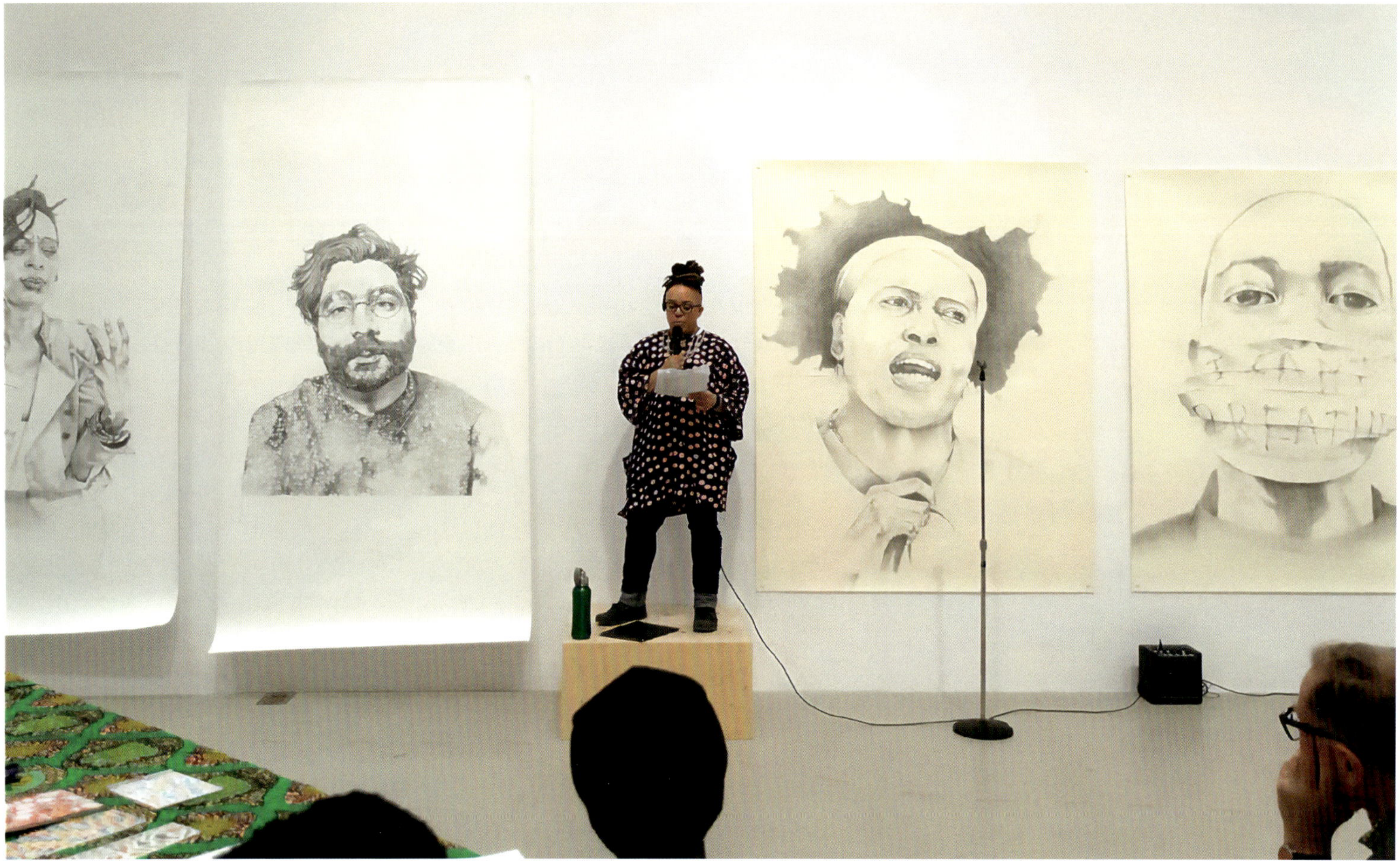

The four artworks in this section highlight and make presentthose people, groups, and communities rendered most vulnerable and targeted through the surveillance tactics of the settler colonial project, which is buttressed by structures of racism, ableism, cisheteropatriarchy, and white supremacy. Not only do they highlight targeted communities, these works also actively involve the viewers, making them participatory community members in their engagement with the work. This aspect should be seen as part of the works' politics and strategies of resistance. Such participatory engagement forecloses the possibility of distanced, disembodied viewing, forcing self-reflection and consideration of each person's complicity and their implicit relationship to the subject at hand. These creative projects humanize and make personal the applications and effects of surveillance, stir relational affinities around the oppressive conditions of surveillance logics, and create space to imagine otherwise. In so doing, they not only engage larger communities, they in fact create

4.9 • Syrus Marcus Ware, *Activist Love Letters*, 2012–ongoing.

community through a shared interaction and engagement with the possibility and power of the people to resist and rethink surveillance operations within the Canadian context.

The artworks in this chapter work together to confront and resist surveillance as a totalizing gaze through which settler colonialism is enacted and reproduced within the Canadian context. They highlight the limits and gaps in surveillance knowledge, and they work to imagine space and possibilities beyond it. Through educating audiences and bringing them together as communities, these works envision creative practice as a vehicle for social change. In conversation with one another, they bring to life Berlant's plan for social change that involves "imagin[ing] the world that you could promise, the world that could be seductive, the world you could induce people to want to leap into."[58] Throughout this chapter, the different projects work to "induce" art audiences to confront, resist, trouble, and move beyond settler colonial structures and surveillance systems that operate in their name to "leap into" new ways of being in the world together.

The three sections here – "Belonging," "Intimacy," and "Community" – organize the artworks into categories that take on different ways of challenging the dehumanizing logics of settler colonial systems and their surveillance apparatuses. In "Belonging," the artworks come together to challenge surveillant policing apparatuses (particularly the passport) that refuse certain forms of belonging to and within the Canadian nation-state, and find space to reclaim attachments outside of these parameters. In "Intimacy," they contest the limits of the surveillance gaze and open up possibilities of communion, vulnerabilities, and togetherness as modes of knowing outside of surveillance data gathering. Finally, in "Community," the creative practices reframe what a for-the-people vision could look like by prioritizing community-based knowledges and practices as central for coming together in compassion and collective power. Ultimately, what underpins the discussion in all of these sections is the element of care. They ask how care can contribute to building togetherness, trust, and compassion, all of which work to counteract and imagine worlds outside the dysregulating and harmful lens of surveillance.

The critical register of these pieces is not a unilateral or parochial reframing of the surveillance narrative of the Canadian nation-state – a state premised on the singular logic of settler colonialism that assesses the value of people according to their position within intersecting systems of oppression, including racism, cisheteropatriarchy, and ableism. While all the works here firmly assert that state and corporate surveillance is a form of social injustice, they do not mount a single form of resistance in response. Rather, they present a multitude of other worlds possible, generated based on local needs and community circumstances, and created by tending to and making space for people as part of a community – of belonging to the land and people around them – that creates intimacies and shared vulnerabilities. However, as pieces like Hameed and Schoepp's *Papers Please!* and Giasson's *VOX* demonstrate, neighbour relations and the watching enacted by individual citizens are not necessarily utopian. They can, in fact, replicate social relations defined by the same systems of inequities and oppression. Yet, as works like Bambitchell's *Border Sounds*, Osman's *Passport*, Dunlop's *A Lover's Geography*, Robertson and White's *Walk the Line* and *Watchers*, and Ware's *Their Borders Crossed Us* and *Activist Love Letters* demonstrate, creative works have a lot of potential to reframe surveillance codes in ways that move beyond flipping the source of social control, focusing, rather, on social relations. They resist and decouple the connections between surveillance viewing and settler colonial viewing, and provide avenues of social and political possibilities outside these terms. Taken together, the artworks throughout this chapter resist and reimagine settler colonial systems in Canada, engage audience-to-community awareness, and develop new spaces of togetherness, care, and compassion to model ethical forms of justice and inclusion.

CONCLUSION

When I started work on this project in 2015, one of my goals was to collect the widest possible array of creative practices that address surveillance in Canada. My preliminary research into the field revealed that creative approaches to the topic of surveillance were notably diverse, and I wanted to bring them together as an archive that catalogues their many different mediums, perspectives, and subject matter. In some ways, I wanted to make sense of this field, this category of "surveillance art" that I have struggled to concretely define. This struggle comes in no small part from the expansive nature of creative engagement with the topic of surveillance, in which categorization feels inadequate. To lump varied approaches together into a tidy category seems a disservice to the complexity and multitudes offered in the meanings and knowledges brought forward by these artworks. While the works all engage with surveillance-related issues and topics, they also produce knowledges beyond this area; they generate multiple layers of complexity not only in their perspectives and interpretations of surveillance but also in so much else.

As I finalized the draft manuscript in spring 2025, I recognized that the ten years since I began this research represents a change in the political landscape that frames the applications and technologies of surveillance. The War on Terror, which was such a dominant paradigm within which surveillance systems were deployed in the preceding twenty-four years, has officially ended, and new contexts and socio-political issues have gained pre-eminence within the public imaginary of how, why, and for whom surveillance structures operate. I was twenty years old in 2001, and my higher education and political cognizance unfolded under the influential rhetoric and policies of the global War on Terror and its increased regimes of surveillance. As a cisgendered, white settler in Canada who is not living with a disability, I am often invisible and unremarkable to the Canadian state's lens of surveillance, and my understanding of surveillance has often felt distanced or theoretical. An element of this research has been learning about and confronting my complicity in the oppressive and inequitable systems that enable surveillance structures. This learning helped me develop a (self-)critical lens through which to understand ongoing iterations of surveillance practices as stemming from the same surveillance logic that, I argue in this book, remains a phenomenon foundational to settler colonial Canada.

I started the discussion in this book with Shaheer Tarar's *Jack Pine* because it seems to me such a keen illustration of the intersections of art, surveillance, and Canadiana. Its reference to Tom Thomson's famous oil painting, its reframing of historical representations, and its exposure of surveillance systems that serve colonial settlerhood bring together the main themes of this book. To me, Tarar's *Jack Pine* is a creative example that highlights the reciprocities between settler colonialism and surveillance, and that seeks to dismantle and denormalize the logics of white settlerhood's claims to control and dominance within Canada. It has been a central aim of this book to shine a light on the intersections between surveillance and settler colonialism, as well as on the perspectives offered by artworks to rethink these intersections, and this work – along with the creative projects that I have detailed throughout this book – demonstrates the complexities offered by artworks on these topics.

Tarar's installation also brings together the historical and ongoing logic of surveillance. By referencing historical representations that foreground questions about for whom the land is marked and claimed within a contemporary context of surveillance, his *Jack Pine* illuminates the pre-2001 roots of the surveillance logic associated with the global War on Terror. This artwork provides an entry point to discuss the expansion of surveillance logic in light of the elevated vigilance that became normalized as essential to conditions of national and global security in the public imaginary after September 2001, to the extent that, in October of that year, former US vice-president Dick Cheney called such vigilance "the new normal."[1] This artwork challenges the accepted idea that surveillance in the Canadian context was new, asserting that the underlying logic of surveillance has long been a vehicle of colonial imaginaries and claims to power that involve containing certain people and communities.

Anthropologist Craig Proulx asserts the idea of "surveillance as a colonial project" and articulates the need to challenge and make "colonial constructions legible and comprehensive for various publics."[2] This book responds to Proulx's call, arguing that creative works provide profound and unique contributions that shape cultural legibility and comprehension of surveillance in Canada in its relation to settler colonialism. Indeed, the artworks in this book demonstrate that surveillance logic is a form of wielding and materializing the systems of power and control that structure and maintain Canada as a setter colonial state and, crucially, that surveillance logic itself predates 2001 and forms an integral part of coloniality.

As projects that are attentive to visibility and visualities, and are practised (intentionally or not) as critical investigations of social processes, artworks that thematize surveillance have constituted unique sites at which to question the technologies themselves, the actors who use them, and the national policies that govern them. Like Torin Monahan, I identify artworks "as being constitutive of the problem spaces they frame and as agential in intervening in those spaces."[3] The problem space in which the artists in this book intervene is, of course, the post-9/11 moment, which sociologist Kevin Haggerty identifies as heralding a new cultural "*zeitgeist* where citizens and officials now assume that greater surveillance is the preferred response to an array of political problems."[4] The creative projects in this book function both as cultural products of and critical interventions in the new cultural zeitgeist, or Cheney's supposed "new normal." They contribute to discussions about how surveillance technologies are used to monitor and track everyday spaces of civilian life, and they provide new and unique ways to think through the political problems of surveillance.

Creative projects can serve as case studies that offer methods for analyzing complex and diffuse ideas about surveillance and coloniality and for rendering them publicly accessible. Surveillance logic in its relationship to settler colonialism is the dominant framework through which I have centred artworks to analyze the creation and maintenance of Canada as a political project, but, in the specific forms of critique brought forward by the creative practices, I have also emphasized how this relationship is bolstered by and contributes to capitalism, imperialism, cisheteropatriarchy, racism, white supremacy, and ableism. The public-facing nature of art provides encounters with audiences in unexpected ways that can challenge, trouble, and reframe their understandings of these systems and their interconnections.

In this way, these creative projects help frame socio-political life. As visual cultural scholar Irit Rogoff asserts, "meaning is not excavated for, but rather … it *takes place* in the present."[5] Here, I understand Rogoff as saying that artworks are participatory in the politics of the everyday – that they do not simply reflect but, rather, produce and shape the contexts with which they engage. These contexts – here related to surveillance in Canada – do not exist only as pre-established norms and cultural beliefs, which artworks then function to confirm or deny. Rather, by facilitating encounters that inform, challenge, reframe, and introduce conceptions of larger socio-political contexts for an audience, artworks engage directly with how people come to understand the world.

In *States of Observance*, my analyses always start with the artwork. Yet I have also been interested in putting these artworks in direct conversation with research in the field of surveillance studies. As a research area dedicated to surveillance systems in a multitude of forms, locations, and historical periods, surveillance studies offers a clear scaffold for the topics, critiques, and themes around surveillance that many of the artworks I encountered identify. To deepen the analysis and examination of these themes and the ways in which artworks engage them, I have also drawn on a wide range of interdisciplinary scholars and writers. So one of my motivations while organizing this book has been to clarify the ways in which art renders aspects of these scholarly conversations "legible," to use Proulx's term. Even simply the act of drawing attention to the hidden ubiquity of surveillance through the creation of artistic interactions is representative of this. I have been interested in how art enters cultural and public spaces to provide new access points of encounter. By consciously curating the art in *States of Observance* in relation to scholarly fields,

I offer art and scholarship in direct relationship and argue that new life and unique perspectives from each area grow from this dialogue.

While my own analyses and the artworks I address in *States of Observance* stop at 2021, the official end of the global War on Terror, conversation and debates about securitization and surveillance logic continue under new circumstances. For instance, the start of the COVID-19 pandemic in 2020 gave new prominence to biometrics and biomedical surveillance, while the work-from-home scenario produced by quarantine and "social distancing" created an increased market for at-home surveillance devices that monitor and track individuals' working and schooling environments.[6] Companies like Proctorio or ProctorU quickly responded to the demand of schooling systems after the global outbreak of COVID-19, as educational institutions were forced to adapt to learning and testing remotely. The technology involves a webcam and microphone that oversee and record students as they take tests (hence the companies' play on the word "proctor"). The recording is then put through an AI system intending to flag for behaviour or activity that could indicate cheating. Students and others have come forward to dispute this surveillance technology, arguing that it inequitably targets disabled students, those who do not have privileged access to work in environments that are perfectly quiet and "neutral," and racialized students whose facial features and skin colour are improperly registered by AI.[7] What the art in this book teaches us is that surveillance logic continues to underpin the political rationality of settler colonialism in Canada and intersectingsystems of oppression and inequity. Certainly, while the discourse and practices of the War on Terror are no longer the dominant framework within which surveillance systems operate and are justified in Canada, surveillance logic continues, responding and adapting to new contexts. As public policy scholar Fahad Ahmad and criminologist Jeffrey Monaghan note, "the landscape of 'war on terror' continues to expand."[8]

As the public understanding of surveillance shifts, creative works addressing surveillance systems respond to and participate in current conversations that highlight, think through, trouble, and contest these different circumstances. Artists in Canada are currently engaging with ongoing surveillance logic in its relation to biometric data and genetics, artificial intelligence, privacy intrusion through home products, drone and satellite imaging, and space colonization missions, as some broad examples. And like the artworks in this book, these creative practices will no doubt illuminate many of the invisible and underreported structures of surveillance logic and draw art audiences in as informed publics. As I have highlighted through the discussion in this book, the practice of art does not simply re-present or reconfirm that which one already knows; it also constitutes an encounter as an opportunity to think and feel otherwise. Thinking through art and creative practices in these terms asserts their capacity to activate, rather than simply respond to, socio-political contexts and issues.

This, to me, is art's work. In this configuration, art is a dynamic practice that makes meanings at the intersection of form, subject matter, artist positionalities, audience subjectivities, and relationships to the socio-political logic of the present historical moment. Surveillance logic is key to framing perceptions of security, safety, inclusion, and exclusion in Canada, and artists and creative practitioners offer profound and unique contributions to shaping cultural understandings of surveillance logic as it was deployed during the period of the War on Terror, including the relationship of this contemporary logic to historical systems of settler colonialism in Canada. In this way, artworks demonstrate how creativity is a political practice with the capacity to explore the contours and contestations of a moment's perception and therefore to contribute to its social, cultural, and political configurations. The thirty-three artworks at the centre of *States of Observance* participate in ongoing conversations and debates about surveillance logic and the regulation of people and communities in relation to systems of oppression and control in the Canadian context from 2001 to 2021. Collectively, they demonstrate why such creative participation remains politically vital.

NOTES

INTRODUCTION

1 I have written about this artwork in more detail elsewhere: see Cahill, "Surveillance Frontierism" and Cahill, "Surveillance Frontierism: Art."

2 I want to note my ambivalence about using the term "Canada." While it is a key point of analysis and context throughout this book, I am aware of the dangers of what critical Muslim studies scholar Shaista Patel identifies as "settler centric moves" – the ways in which white settler scholars orient critiques that reify the permanence of white settler colonialism and its violences (see Patel, "Talking Complicity"). However, particularly as a white settler scholar, I am also conflicted about other naming options for the geographic area of northern North America. So, in this book, I hesitantly rely on the term "Canada" to interrogate the narratives, structures, and ideologies that the nation-state represents and perpetuates. And, rather than treating Canada as a neutral or benign geographic or political entity, I approach it as a construct embedded with colonial histories, racialized power relations, and socio-economic hierarchies. By critically engaging with Canada as a category, I aim to challenge the dominant national mythologies that frame it as a peaceful, multicultural, and progressive country, and instead expose the exclusions and violences that underpin its formation and ongoing governance. I use this term "Canada" with caution and awareness of its inadequacy within the context of the goals of my project.

3 The catalogue for this exhibition, dated 6 November–7 December 2019, was created in collaboration with Philip Ocampo and can be found on Sophia Oppel's website, at https://sophia-oppel-art.format.com/collaborations/scaffolds-i-can-no-longer-see.

4 Shaheer Tarar, interview by the author, Zoom, 25 February 2021.

5 For more information, see the website Art and Surveillance, at www.artandsurveillance.com.

6 While this book recognizes multiple, intersecting systems of oppression – such as ableism, racism, sexism, and cisheteronormativity – as intrinsic to the settler colonial project, it does not address each in equal depth, nor does it provide sustained analysis of all of these systems. This limitation reflects the specific scope and focus of the study and is not a dismissal of the significance of any of these systems. Rather, it points to the necessity for further work that takes up these structures in more sustained and dedicated ways.

7 Ożóg, "Surveilling the Surveillance Society," 96.

8 Monahan, "Ways of Being Seen," 565.

9 Brighenti, "Artveillance," 137.

10 Brighenti, "Artveillance," 137.

11 Morrison, *Discipline and Desire*, 7.

12 Monahan, *Crisis Vision*, 9.

13 Bruno, Barreto, and Szafir, "Surveillance Aesthetics," 83.

14 Kafer, "Reimagining Resistance," 230.

15 For readings on the politics of visuality, see Fleetwood, *Troubling Vision*; Mirzoeff, "On Visuality"; Mirzoeff, *The Right to Look*; Hal Foster, ed., *Vision and Visuality*; and Jay and Ramaswamy, eds., *Empires of Vision*.

16 Mirzoeff, "On Visuality," 67.

17 Mitchell, "Showing Seeing," 171.

18 Lauvin, "Neutralizing the Surveillant Gaze," para. 3.

19 Fleetwood, *Troubling Vision*, 7.

20 Finn, "Seeing Surveillantly," 67.

21 Dewey, *Art as Experience*, 3.

22 Murray-Lichtman and Elkassem, "Academic Voyeurism," 181.

23 Andrejevic, "Foreword," x.

24 Lyon, *Surveillance Society*, 2.
25 Haggerty and Ericson, "New Politics," 3.
26 Bennett, Haggerty, Lyon, and Steeves, eds., *Transparent Lives*, vii. The New Transparency Project operated from 2008 to 2015 as a research team dedicated to the study of surveillance and governance. It involved over thirty-three scholars from around the world. Its main website is no longer active, but more information on the project is available at https://dcssproject.net/the-new-transparency/index.html.
27 Monahan, *Crisis Vision*, 6.
28 Cahill, "Surveillance Frontierism."
29 Nunn, "Toxic Encounters," 1337.
30 Wolfe, "Settler Colonialism," 388.
31 Monahan, *Crisis Vision*, 139.
32 Glen Coulthard, "For Our Nations to Live, Capitalism Must Die," *Unsettling America: Decolonization in Theory & Practice* (blog), 5 November 2013, https://unsettlingamerica.wordpress.com/2013/11/05/for-our-nations-to-live-capitalism-must-die/.
33 See Deleuze, "Postscript on the Societies of Control."
34 Majid, "Spectres of the Moor," in *We Are All Moors*, 1–29.
35 Patel, Moussa, and Upadhyay, "Complicities, Connections and Struggles," 9.
36 Bennett, Haggerty, Lyon, and Steeves, eds., *Transparent Lives*, 10.
37 McDonald, *Identity Discourses*, 1.
38 For a critical analysis of this bill in relation to Islamophobia in Canada, see Patel, "Anti-Terrorism Act."
39 Forcese and Roach, "Bill C-51"; Hayden Watters, "C-51, Controversial Anti-Terrorism Bill, Is Now Law. So, What Changes?," CBC News, 18 June 2015, http://www.cbc.ca/news/politics/c-51-controversial-anti-terrorism-bill-is-now-law-so-what-changes-1.3108608.
40 "Bill C-59: Mass Surveillance and Cyber Powers," International Civil Liberties Monitoring Group, accessed 15 June 2024, https://iclmg.ca/issues/bill-c-59-the-national-security-act-of-2017/bill-c-59s-mass-surveillance-and-cyber-powers/.
41 "Five Eyes," Privacy International, accessed 15 June 2024, https://privacyinternational.org/learn/five-eyes.
42 James Cudmore, "Canada to Pull Out of Key NATO Air Defence Program," CBC News, 9 June 2011, http://www.cbc.ca/news/politics/story/2011/06/09/pol-nato-awacs-canada-mackay.html; Laura Payton, "Canada Pulls Out of NATO Surveillance Project," CBC News, 9 Feb 2012, http://www.cbc.ca/news/politics/story/2012/02/09/pol-canada-nato-alliance-ground-surveillance.html; David Pugliese, "Canada Pulls Out of NATO Airborne Surveillance Program to Save $90M," *National Post*, 19 March 2012, http://news.nationalpost.com/2012/03/17/canada-pulls-out-of-nato-airborne-surveillance-programs-to-save-90m/.
43 Lee Berthiaume, "Trudeau Heads to NATO Summit with Commitments, but Will Face Questions," *iPolitics*, 7 July 2016, https://ipolitics.ca/2016/07/07/trudeau-heads-to-nato-summit-with-commitments-but-will-face-questions/; "Canada Makes Commitment to NATO Defence and Deterrence Measures," Prime Minister of Canada website, 8 July 2016, http://pm.gc.ca/eng/news/2016/07/08/canada-makes-commitment-nato-defence-and-deterrence-measures.
44 Lyon, "9/11, Synopticon, and Scopophilia," 39.
45 Hier and Greenberg, "Politics of Surveillance," 28.
46 For instance, see Janes and Sandell, eds., *Museum Activism*; O'Neill, *Culture of Curating*; and Rudd and Sedgwick, eds., *Issues in Curating*.
47 Kim, "Politics of Citation," 7.
48 Tuck and Yang, "Decolonization Is Not a Metaphor," 21.
49 Morton, "Introduction," 8.

CHAPTER ONE

1 See, for example, Finn, "Surveillance Studies and Visual Art"; Lyon, *Surveillance after September 11*; Lyon, *Surveillance Studies*; and van der Meulen and Heynan, eds., *Expanding the Gaze*.
2 Sturken and Cartwright, *Practices of Looking*, 103.
3 See, for example, Gandy, "Data Mining and Surveillance"; Kinsman, Buse, and Steedman, eds., *Whose National Security?*; Sorkin, ed., *Indefensible Space*; Zureik et al., eds, *Surveillance, Privacy, and the Globalization of Personal Information*.
4 Foucault, *Discipline and Punish*.
5 Hier, "Probing the Surveillant Assemblage," 403; Lyon, *Surveillance Studies*, 57.
6 See Matthew Allen, "See You in the City!"; and Sherman, "Polyopticon."
7 See Wood, "Foucault and Panopticonism Revisited"; and Yar, "Panoptic Power."
8 Browne, *Dark Matters*, 24.
9 Robertson, "'Try to Walk,'" 24.
10 Rice, Dion, Fowlie, and Mündel, "Re/turning the Gaze," 639.
11 Weinberg, "Surveillance as Paradox."
12 Griffiths, "The White Gaze," 415.
13 Sa'di, "Colonialism and Surveillance," 153.
14 Nakamura, "Afterword," 221. Emphasis in the original.
15 Finn, "Seeing Surveillantly," 67.
16 Finn, "Seeing Surveillantly," 67.

17 bell hooks, *Black Looks*, 6.
18 Allen, "Sorting Daemons," 26.
19 Tarah Hogue, "Unsettled Sites: Marian Penner Bancroft, Wanda Nanibush, and Tania Willard," exhibition notes, SFU Gallery, accessed 21 June 2024, https://www.sfu.ca/galleries/sfu-gallery/past/UnsettledSites.html.
20 "Conversation – Tarah Hogue, Marian Penner Bancroft and Tania Willard: Unsettled Sites," 7 May 2016, YouTube video, 26:24, https://www.youtube.com/watch?v=4QEfVv1iXMg&t=361s.
21 The website and linked security camera images can be found on the "Cams" page of the Beausoleil First Nation website, at http://www.chimnissing.ca/index.html.
22 Boisjoly and Gray, "Unsettled Sites."
23 See, for instance, Crosby and Monaghan, *Policing Indigenous Movements*, and Keith D. Smith, *Liberalism, Surveillance, and Resistance*.
24 Boisjoly and Gray, "Unsettled Sites."
25 Antonia Hirsch, "Double Blind," artist's website, accessed 17 June 2024, http://antoniahirsch.com/works/double-blind/.
26 Kwan, "About the Work."
27 Allen, "Sorting Daemons," 12.
28 Jill Magid's *Evidence Locker* (2004) is another artwork that highlights the spaces of unknowing inherent to seeing through the surveillance gaze.
29 See Lyon, "Surveillance and the Eye of God."
30 Lewis, *Under Surveillance*, 226.
31 "and all watched over by machines of loving grace," artist Donna Szoke's website, accessed 17 June 2024, https://donnaszoke.com/?projects=all-watched-over.
32 "And all watched over," Szoke website, https://donnaszoke.com/?projects=all-watched-over.
33 Exploring the surveillance gaze through asserting the centrality of the human body – particularly the eyes – as the root of that gaze is not unique to the fine art context. Cultural products outside the fine arts also play with the technologies of vision and the human eye. One notable example is the work of engineer Steve Mann, who pioneered prosthetic surveillance technology that allows the wearer to track people and activities from their own point of view.
34 Moser, "Jennifer Marman."
35 On Art Souterrain's official website, the organization describes TOPO and its mandate thusly: "The Montreal artist-run centre TOPO is a laboratory for digital writings and creations for web, performance, and installation spaces. Its mandate is to incubate, produce, and circulate original multimedia artworks that explore interdisciplinary and intercultural hybridizations in the digital arts." Art Souterrain, "TOPO," Art Souterrain website, accessed June 2018, http://www.agencetopo.qc.ca/wp/en/.
36 Art Souterrain, "Description," Art Souterrain website, accessed June 2018, http://2015.artsouterrain.com/fr/pat-badani/.
37 Monahan, "Ways of Being Seen," 561.
38 Gavin J.D. Smith, "Surveillance, Data, and Embodiment," 109.
39 To view images of the artworks and installation from this exhibition, please visit the *Mi'kwite'tmn* website: https://mikwitetmn.ca/.
40 "Exhibition: Mi'kwite'tmn (Do You Remember)," Mi'kwite'tmn exhibition website, accessed 16 May 2024, https://mikwitetmn.ca/.
41 Johnson, "Ursula Johnson Q&A."
42 Johnson, "Ursula Johnson Q&A."
43 Dickenson, "Processing," 125. Emphasis in the original.
44 Significantly, later iterations of this touring exhibition seem to have revised this surveillance aspect. Rather than cameras mounted in the gallery space monitoring the audience, Johnson installed an in-counter barcode scanner – similar to what would be in most grocery stores – so that visitors to the space could select objects and scan them to get more information and details about the work.
45 Johnson, "Ursula Johnson Q&A."
46 "Interactive Installations: Gathering (2004)," David Rokeby's website, accessed 23 May 2024, http://www.davidrokeby.com/gathering.html.
47 Buolamwini and Gebru, "Gender Shades."
48 "Overview," Gender Shades website, accessed 17 June 2024, http://gendershades.org/overview.html.
49 Raji et al., "Saving Face," 145.
50 Hier, "Probing the Surveillant Assemblage," 400.
51 Benjamin, *Race after Technology*, 2.
52 Noble, *Algorithms of Oppression*, 1.
53 Monaghan, "Settler Governmentality," 488.

CHAPTER TWO

1 See Bennett, Haggerty, Lyon, and Steeves, eds., *Transparent Lives* for a thorough analysis of the various trends and patterns in contemporary Canadian surveillance systems. The authors of that book are part of a larger research group called the New Transparency Project. Project information available here: https://dcssproject.net/the-new-transparency/index.html.
2 This expanded state of surveillance was labelled "the new normal" in October 2001 by former US vice-president Dick Cheney. Quoted in Conner, "The New Normal," 12.

3 Haggerty and Ericson, "New Politics," 6.
4 Monahan, "The Right to Hide?," 160.
5 The Coalition of the Willing in the context of the War on Terror refers to the international alliance of military forces that came together as part of the American-led invasion of Iraq in 2003.
6 Patel, "Anti-Terrorism Act," 273.
7 Fiske, "Surveilling the City," 69.
8 Ahmad and Monaghan, "Socializing the High Policing Métier," 248. The authors reference Brodeur, "High Policing."
9 Kaba, *We Do This*, 256.
10 Maynard, *Policing Black Lives*, 7.
11 Maynard, *Policing Black Lives*, 87.
12 Monaghan, "Security Traps."
13 Monaghan and Crosby, *Policing Indigenous Movements*. Further information about this project and book is available at Policing Indigenous Movements: Dissent and the Security State, https://policingindigenousmovements.ca/introduction/. Project SITKA is an RCMP-led operation, launched in 2014, that targeted Indigenous rights activists. It was developed in the wake of the 2013 Idle No More protests and included a list of eighty-nine people who were labelled as security threats. For more details on this project, see Jorge Barrera, "RCMP Intelligence Centre Compiled List of 89 Indigenous Rights Activists Considered 'Threats,'" APTN National News, 8 November 2016, https://www.aptnnews.ca/national-news/rcmp-intelligence-centre-compiled-list-of-89-indigenous-rights-activists-considered-threats/.
14 Monahan, "The Right to Hide."
15 Haggerty, Wilson, and Smith, "Theorizing Surveillance," 233.
16 Newell, "Context, Visibility, and Control," 60.
17 Egbert and Mann, "Discrimination in Predictive Policing," 27.
18 Egbert and Mann, "Discrimination in Predictive Policing," 27.
19 Edwin Janzen, "Inspirations Got to Come from Somewhere," Conspiracy Case website, August 2012, https://conspiracycase.wordpress.com/2012/08/.
20 "Conspiracy Case: A Fantasy of Total Surveillance by Edwin Janzen," Conspiracy Case website, 2012, http://edwin-janzen.weebly.com/conspiracy-case.html.
21 Conspiracy Case website, http://edwin-janzen.weebly.com/conspiracy-case.html.
22 Michael Lefebure, *Criminal Conspiracy: Issues and Complexity*, Canadian Legal Information Institute, May 2023, CanLIIDocs 930, https://canlii.ca/t/7n426.
23 Smith, "Sociable Laughter," 3.
24 See, for example, Amnesty International, "Carding and Anti-Black Racism in Canada," *Amnesty International* (blog), 23 August 2019, https://www.amnesty.ca/blog/carding-canada/; and Ontario Human Rights Commission, "Stop and Search."
25 Ontario Human Rights Commission, "Stop and Search."
26 Akshay Kulkarni, "Data Shows Indigenous, Black People Have More Police Interactions. Advocates Say It Reflects Systemic Racism," CBC News, 12 February 2023, https://www.cbc.ca/news/canada/british-columbia/bc-systemic-racism-police-1.6744961.
27 Cole, *The Skin We're In*.
28 Maynard, *Policing Black Lives*.
29 This artwork was exhibited in other locales, including at the Science Gallery, Dublin, Ireland. These earlier iterations used a different training corpus, which later became refined as Agent Baines in the Calgary version of the piece.
30 Gradecki and Curry, "Crowd-Sourced Intelligence Agency."
31 Gradecki and Curry, "Crowd-Sourced Intelligence Agency."
32 Zine, *Under Siege*, 3.
33 Gradecki and Curry, "Crowd-Sourced Intelligence Agency."
34 Page and Jones, "Weaponizing Neutrality," 172, 171.
35 Stahl and Kaempf, "Sousveilling," 338.
36 Kaba, *We Do This*, 139.
37 "Twenty Four Seven/365," Aljumaine Gayle website, accessed 17 June 2024, https://aljumaine.com/twentyfourseven.html.
38 Kate Schneider, "Use of Controversial Surveillance Technology Demonstrates the Need to Limit Police Power," CBC Opinion, 8 January 2022, https://www.cbc.ca/news/opinion/opinion-police-facial-recognition-technology-clearview-ai-1.6306357; Nicole Brockbank, "Toronto Police Used Clearview AI Facial Recognition Software in 84 Investigations," CBC News, 23 December 2021, https://www.cbc.ca/news/canada/toronto/toronto-police-report-clearview-ai-1.6295295.
39 See, for example, Shapiro, "Predictive Policing for Reform?"; Browning and Arrigo, "Stop and Risk"; Duarte, "Making of Crime Predictions"; and DaViera, Uriostegui, Gottlieb, and Onyeka, "Risk, Race, and Predictive Policing."
40 Maynard, *Policing Black Lives*, 88.
41 Browne, *Dark Matters*, 16, quoted in *Twenty Four Seven/365*, artist's Aljumaine Gayle website, accessed 16 June 2024, https://aljumaine.com/twentyfourseven.html.
42 Gandy, *Panoptic Sort*.
43 Colin Bennett, "In Defense of Privacy," 490.
44 "URME Surveillance," Leonardo Selvaggio website, accessed 14 June 2024, http://leoselvaggio.com/urmesurveillance.
45 Monahan, "The Right to Hide?," 159.

46 Monahan, "The Right to Hide?," 164–5.
47 The studio was renamed Ada X in 2020. Prior to this, it was called Studio XX.
48 "Stranger Visions," Heather Dewey-Hagborg website, accessed 14 June 2024, https://deweyhagborg.com/projects/stranger-visions.
49 Stroud, "The Most Advanced Police Sketch."
50 Queirós, "Visibilities and Invisibilities," 1.
51 "A New Technology That Focuses on DNA May Help Police Solve Cold Cases," Citytv, City News Toronto, June 2014, https://players.brightcove.net/593482582001/default_default/index.html?videoId=3415733834001.
52 Andrew Pollack, "Building a Face, and a Case, on DNA," *New York Times*, 23 February 2015, https://www.nytimes.com/2015/02/24/science/building-face-and-a-case-on-dna.html.
53 Sarah Rieger, "Police Sketch Created with DNA Technology Is Potentially Useless or Even Misleading, Says Scientist," CBC News, 22 February 2018, https://www.cbc.ca/news/canada/calgary/police-sketch-dna-1.4547272.
54 Caley Gibson, "Edmonton Police Apologize after Concerns Raised over Release of DNA Phenotyping Composite Sketch," Global News, 6 October 2022, https://globalnews.ca/news/9181617/edmonton-police-apology-dna-phenotyping-composite-sketch/.
55 Dewey-Hagborg, "Sci-Fi Drama."
56 Hopman and M'charek, "Facing the Unknown Suspect," 439.
57 Dewey-Hagborg, "Sci-Fi Drama."
58 Tremblay, "*Stranger Visions*," 45.
59 Barnard-Wills and Wells, "Surveillance, Technology," 229.
60 Lyon, "Surveillance Technology," 163.
61 Vukov and Sheller, "Border Work," 230.
62 Kate McQuillen quoted in June Chua, "Artist Probes Surveillance and Limits of Privacy," *Rabble.ca*, 7 August 2013, https://rabble.ca/columnists/artist-probes-surveillance-and-limits-privacy/.
63 Browne, *Dark Matters*, 9.
64 van der Ploeg, "The Body as Data," 179.
65 Lyon, "Under My Skin," 305.
66 Glasbeek, Alam, and Roots, "Seeing and Not-Seeing," 329.

CHAPTER THREE

1 Anouar Majid's book *We Are All Moors* discusses the historical frameworks of contemporary surveillance. As he states, "the rise of a surveillance society in the last few decades and the steady encroachments on the rights to privacy as necessary undertakings to protect nations against terrorists and immigrants are elements of a story that was promoted by the Inquisition in early modern Spain to protect Iberia's cultural purity" (6–7).
2 Haggerty and Trottier, "Surveillance and/of Nature," 401.
3 Cronon, "The Problem with Wilderness," 7.
4 For more discussion on these intersections, see, for instance, Maynard et al., "Every Day" and Buenavista et al., "A Praxis."
5 Moreton-Robinson, *White Possessive*, xii.
6 Gebhard, McLean, and St. Denis, eds., *White Benevolence*, 2.
7 Browne, *Dark Matters*, 17.
8 Wolfe, "Settler Colonialism," 388.
9 Robertson, "Notes to a Nation," 65.
10 Coulthard, *Red Skin, White Masks*, 14.
11 McKay, *Picturing the Land*, 5.
12 For further analysis that connects the representational and the material in the context of railway construction, colonial dispossession, and gendered, racialized, classed, sexualized, able-bodied, and citizen subjectivities, see Valadares, "Uneven Mobilities."
13 Bear Nicholas, "Role of Colonial Artists," 28–9.
14 Sa'di, "Colonialism and Surveillance," 154.
15 Edney, *Mapping an Empire*, 1.
16 Bennett, Greaves, Riedlsperger, and Botella, "Articulating the Arctic," 630.
17 Goeman, "Tools of a Cartographic Poet," 89.
18 "Satellites," Hiba Ali website, accessed 14 May 2024, https://hibaali.info/projects/satellites.
19 This work was also exhibited in Austin, Texas, and Dubai in 2017.
20 Gál, "Interstellar Railroad."
21 Jon Rafman, "IMG MGMT: The Nine Eyes of Google Street View," ArtFCity website, 12 August 2009, http://artfcity.com/2009/08/12/img-mgmt-the-nine-eyes-of-google-street-view/.
22 Ingraham and Rowland, "Performing Imperceptibility," 211.
23 Ingraham and Rowland, "Performing Imperceptibility," 211.
24 Farman, "Mapping the Digital Empire," 877.
25 "Our Approach to Search," Google Search, accessed 17 June 2024, https://www.google.com/search/howsearchworks/our-approach/.
26 Rafman, "IMG MGMT."
27 Gibson, "Narrative Hunger," 256.
28 Zuboff, *Surveillance Capitalism*, 141.
29 Benson et al., "Mapping the Spatial Politics," 1.
30 Mesquita, "Counter-Cartographies," 26.
31 Andrews, "Hiding or Hidden," 5.
32 *Mamook Ipsoot*, grunt gallery website, October 2013, https://grunt.ca/exhibitions/mamook-ipsoot/.

33 In the discussion of the production of the pieces, project coordinator Jolene Andrews highlights the sociality and collaboration of the artists with each other and the works. While each artist was responsible for producing their own image, Andrew emphasizes how the process very much created a community of those involved. Andrews, "Hiding or Hidden," 9.

34 Jaden Jugovic, quoted in grunt gallery, *Mamook Ipsoot*, 18; Laticia Baptiste, quoted in grunt gallery, *Mamook Ipsoot*, 17.

35 Carson Pechawis, quoted in grunt gallery, *Mamook Ipsoot*, 19.

36 Johns, "Indigenous World-Building."

37 Waldron, "Mapping Racial Geographies," 2.

38 Moreton-Robinson, *White Possessive*, xiii.

39 Goldstein, "On the Reproduction of Race," 42.

40 Moreton-Robinson, *White Possessive*, xix.

41 McKnight, "Creative Research Methodologies," 154.

42 Stéfy McKnight, "Artist Interview: Stéfy McKnight, Creator of 'Hunting for Prey,'" interview by Torin Monahan, *Medium*, 10 September 2018, https://medium.com/surveillance-and-society/artist-interview-stefy-mcknight-creator-of-hunting-for-prey-7a59ae62e227.

43 Haggerty and Trottier, "Surveillance and/of Nature," 401.

44 Wakefield, "Public Surveillance Functions," 529.

45 Cyphers and Gebhart, *Behind the One-Way Mirror*, 4.

46 Moreton-Robinson, *White Possessive*, xxiii.

47 Moreton-Robinson, *White Possessive*, xxiii.

48 Robertson, "Notes to a Nation," 66.

49 This piece is a site-specific work that Rokeby recreated in several different locations. He also developed a newer incarnation of the piece, *Machine for Taking Time (Boul. Saint-Laurent)* (2006–07), commissioned by the Daniel Langlois Foundation. More information on this latter piece, including detailed interviews with the artist, can be found on the foundation's website: http://www.fondation-langlois.org/e-art/e/machine-for-taking-time.html.

50 "Machine for Taking Time, David Rokeby," DOCAM: Documentation and Conservation of the Media Arts Heritage, accessed 17 June 2024, https://www.docam.ca/en/common-problems/machine-for-taking-time-david-rokeby.html.

51 "Video Installations: Machine for Taking Time (2001–4)," David Rokeby website, accessed 23 May 2024, http://www.davidrokeby.com/machine.html.

52 Simpson, "Whither Settler Colonialism?," 439.

53 Fung, "Review of David Rokeby."

54 Bouchard, "Observation Instruments," 69.

55 "Pierre Tremblay – Meta Incognita: Summer Variations," The Photographers' Gallery website, accessed 17 June 2024, https://thephotographersgallery.org.uk/whats-on/pierre-tremblay-meta-incognita-summer-variations.

56 Bouchard, "Observation Instruments."

57 Bouchard, quoted in Loft, "Reconstructions and Digital Landscapes," 3.

58 Loft, "Reconstructions and Digital Landscapes," 1.

59 "Philip Cheung: Arctic Front," Circuit Gallery website, accessed 17 June 2024, http://www.circuitgallery.com/exhibitions/philip-cheung-arctic-front/.

60 Ellis Quinn, "The Canadian Rangers at 75: The Eyes and Ears of the North," Eye on the Arctic, Radio Canada International, 2 May 2022, https://www.rcinet.ca/eye-on-the-arctic/2022/05/02/the-canadian-rangers-at-75-the-eyes-and-ears-of-the-north/.

61 "Philip Cheung," Circuit Gallery website, http://www.circuitgallery.com/exhibitions/philip-cheung-arctic-front/.

62 Government of Canada, "About the Canadian Rangers," Department of National Defence and Canadian Armed Forces Ombudsman, last modified 14 June 2023, https://www.canada.ca/en/ombudsman-national-defence-forces/education-information/caf-members/career/canadian-rangers.html.

63 Mike Blanchfield, "House Panel Urges Canada to Work with NATO to Protect Arctic Sovereignty," Global News, 10 April 2019, https://globalnews.ca/news/5153484/canada-arctic-sovereignty-nato-russia/.

64 Sara Matthews, "Watching the Watchers," Exhibition essay, Circuit Gallery webiste, accessed 17 June 2024, http://www.circuitgallery.com/exhibitions/philip-cheung-arctic-front/sara-matthews-arctic-front-essay/.

65 Ogasawara, "Mainstreaming Colonial Experiences," 727.

CHAPTER FOUR

1 Mirzoeff, "Right to Look," 477.

2 Robé, *Abolishing Surveillance*, 127.

3 Keene and Hitch, "Drone Warriors." An art exhibition of some of the images produced by this group was curated by Keene and Hitch and hosted by the Haffenreffer Museum of Anthropology, Brown University, Providence, Rhode Island, from May 2018 to December 2019. Information on this exhibition can be found on the Haffenreffer Museum of Anthropology webpage, at https://hma.brown.edu/exhibits/past-exhibits/drone-warriors-art-surveillance-and-resistance-standing-rock.

4 Sholette, *Art of Activism*, 10.
5 Robertson, *Tear Gas Epiphanies*, 14.
6 Stevens and Lossau, "Framing Art," 2.
7 I discuss some of the specific Canadian laws and policies developed in the wake of 9/11 in the introduction to this book.
8 Simpson, "The State Is a Man."
9 Gill, "Surveillance Is a Feminist Issue," 150.
10 Gradecki and Curry, "Crowd-Sourced Intelligence Agency."
11 Maynard, *Policing Black Lives*, 182.
12 Sharma, "Racism," 99; emphasis in the original.
13 Hameed, "Borders in the City," 121.
14 Hameed, "Borders in the City," 133, 121.
15 Hameed, "Borders in the City," 132.
16 Hameed, "Borders in the City," 132–3.
17 Hameed, "Borders in the City," 131–2.
18 The commission's co-chairs were Charles Taylor, a political scientist and philosopher and professor emeritus at McGill University, and Gérard Bouchard, a historian and sociologist affiliated with l'Université du Québec à Chicoutimi. Mahrouse, "'Reasonable Accommodation,'" 86.
19 Mahrouse, "'Reasonable Accommodation,'" 86.
20 Zine, *Under Siege*, 6.
21 Gérard Bouchard and Taylor, *Building the Future*, 22. As to what these efforts look like, the report's recommendations are quite limited. The researchers assert the need for "interculturalism," which they describe as the need "to reconcile ethnocultural diversity with the continuity of the French-speaking core and the preservation of the social link. It thus affords security to Quebecers of French-Canadian origin and to ethnocultural minorities and protects the rights of all in keeping with the liberal tradition" (19). So too, they emphasize the responsibilities of ethnic minorities to assert themselves within French-Canadian life: "Members of the ethnic minorities must do more to make themselves seen and heard" (250). So, while the report asserts the need for more inclusion, that inclusion is defined in relation to maintaining and centring Québécois traditions.
22 Bambitchell, "Sounding the Border," 88.
23 Browne, *Dark Matters*, 134.
24 Bambitchell, "Sounding the Border," 82. Original is written in italics.
25 Bambitchell, "Sounding the Border," 85.
26 Bambitchell, "Sounding the Border," 86.
27 Puar, *Terrorist Assemblages*, 2.
28 "About," Abdi Osman website, accessed 18 June 2024, http://abdiosman.com/about/.
29 Sharpe, "Variations in Black Queer Worlds."
30 Walker, "Variations in Black, Queer, and Otherwise."
31 Walker, "Variations in Black, Queer, and Otherwise."
32 Sharpe, "Variations in Black Queer Worlds."
33 Browne, *Dark Matters*, 143.
34 See, for example, Wallace, "Our Lives"; and Willett and Smaal, eds., *Intimacy, Violence and Activism*. Also see the exhibition *A Different Kind of Intimacy: Queer and Radical Performance at the Walker, 1990–1995*, curated by Gwyneth Shanks at the Walker Art Center, Minneapolis, in 2018, https://walkerart.org/press-releases/2018/walker-art-center-presents-a-different-kind-of-intimacy-queer-and-radical-performance-at-the-walker-1990-1995.
35 Lauren Berlant, quoted in Davis and Sarlin, "'On the Risk."
36 Garcia-Salazar, Loshkajian, Cahn, and Manis, *Spy Next Door*.
37 See Burchell, Gordon, and Miller, eds., *Foucault Effect*.
38 An interesting creative and critical online work that engages with these larger issues is *White Collar Crime Risk Zones*, developed by Brian Clifton, Sam Lavigne, and Francis Tseng for *New Inquiry Magazine* 59 (March 2017). It can be accessed on the New Inquiry website, at https://whitecollar.thenewinquiry.com/.
39 Godfrey, "Colour, Abstraction and Queerness."
40 McGrath, *Loving Big Brother*, 20–1.
41 Bishop, "Participation and Spectacle," 36.
42 Browne, *Dark Matters*, 8.
43 "Walk the Line (2014–2015)," Genevieve Robertson website, accessed 19 June 2024, https://www.genevieverobertson.com/#/walk-the-line/.
44 "Walk the Line – 2015, 2015," Jay White website, accessed 19 June 2024, https://draworbedrawn.com/Walk-The-Line-2014-2015.
45 Martin Lukacs, "Indigenous Rights a 'Serious Obstacle' to Kinder Morgan Pipeline, Report Says," *Guardian*, 16 October 2017, https://www.theguardian.com/environment/true-north/2017/oct/16/indigenous-rights-serious-obstacle-to-kinder-morgan-pipeline-report-says.
46 The first iteration of Robertson and White's *Walk the Line* project took place on 14 December 2014.
47 "Watchers (2015)," Genevieve Robertson website, accessed 19 June 2024, https://www.genevieverobertson.com/watchers.
48 "Watchers," Robertson website, https://www.genevieverobertson.com/watchers
49 For media coverage of some of these arrests, see, for example, "5 Arrested after Standing in Way of Trans Mountain Pipeline Construction in B.C. Interior," CBC News, 16 October 2020, https://www.cbc.ca/news/canada/british-columbia/trans-mountain-pipeline-kamloops-thompson-river-secwepemc-1.5765885;

Chad Pawson, "Pipeline Protesters Predict Number of Arrests over Trans Mountain Will Eclipse War in the Woods," CBC News, 16 July 2018, https://www.cbc.ca/news/canada/british-columbia/trans-mountain-pipeline-expansion-arrests-1.4747601; Jesse Ferreras, "Burnaby RCMP Arrest 3 Protesters Who Zap-Strapped Themselves to Kinder Morgan Gate," Global News, 12 April 2018, https://globalnews.ca/news/4140977/kinder-morgan-protest-artist-arrests/; Maya Laframboise, "To Stop a Pipeline, a Three-Week Jail Sentence Was a Small Price to Pay," *The Breach*, 15 August 2023, https://breachmedia.ca/to-stop-a-pipeline-a-three-week-jail-sentence-was-a-small-price-to-pay/.

50 For media coverage on Indigenous-led actions and arrests, see, for example, "Indigenous Elders Arrested at Kinder Morgan Protest," Last Real Indians website, 23 August 2018, https://lastrealindians.com/native-news-desk/2018/8/23/aug-23-2018-indigenous-elders-arrested-at-kinder-morgan-protest; Brett Forester, "RCMP Cited Standing Rock Protests as 'Evidence' to Create Controversial B.C. Unit," CBC News, 3 August 2023, https://www.cbc.ca/news/indigenous/rcmp-cirg-standing-rock-documents-1.6925906; Kamyar Razavi, "Indigenous Rights Collide with $35B Western Canada Pipeline Expansion," Global News, 22 November 2023, https://globalnews.ca/news/10103531/indigenous-rights-collide-with-35b-western-canada-pipeline-expansion/.

51 For media coverage on this issue, see "Why Building a Pipeline on Indigenous Land Is Complicated Even if You Own It," APTN National News, 5 July 2018, https://www.aptnnews.ca/national-news/why-building-a-pipeline-on-indigenous-land-is-complicated-even-if-you-own-it/. The Kinder Morgan pipeline was bought for $4.5 billion by the Canadian federal government in 2018. Since that time, the cost of the project has skyrocketed to $21.4 billion. Brieanna Charlebois, "Indigenous Leaders, Protesters Gather in Vancouver to Oppose Trans Mountain Pipeline," Global News, 9 April 2022, https://globalnews.ca/news/8749186/indigenous-leaders-protesters-gather-in-vancouver-to-oppose-trans-mountain-pipeline/.

52 Mitra, ed., *Border Cultures*, 134; "Activist Portrait Series," Syrus Marcus Ware website, accessed 19 June 2024, https://www.syrusmarcusware.com/art/activist-portrait-series. More information about these activists and the work they are doing is included in the *Border Cultures* exhibition catalogue.

53 Mitra, ed., *Border Cultures*, 134.

54 Mitra, ed., *Border Cultures*, 134.

55 "In the Studio with Syrus Marcus Ware," Canadian Art, 19 June 2018, Vimeo video, 3:16, https://vimeo.com/275909676.

56 Syrus Marcus Ware, "Queer (Self) Portraits: Syrus Marcus Ware," CBC Arts, 20 June 2017, YouTube video, 3:42, https://www.youtube.com/watch?v=MMsr6ukWKBA.

57 "Activist Love Letters," Syrus Marcus Ware website, 19 June 2024, https://www.syrusmarcusware.com/art/activist-love-letters.

58 Lauren Berlant, quoted in Davis and Sarlin, "'On the Risk.'"

CONCLUSION

1 Conner, "New Normal," 12.

2 Proulx, "Colonizing Surveillance" 95.

3 Monahan, *Crisis Vision*, 143.

4 Haggerty, "Foreword," ix.

5 Rogoff, "Smuggling"; emphasis in the original.

6 There have been a number of fascinating critiques of expanded surveillance systems in the context of COVID-19. See, for example, Todd Feathers, "An Exam Surveillance Company Is Trying to Silence Critics with Lawsuits," *VICE*, 21 October 2020, https://www.vice.com/en/article/7k9zjy/an-exam-surveillance-company-is-trying-to-silence-critics-with-lawsuits; Robyn Maynard and Andrea J. Ritchie, "Black Communities Need Support, Not a Coronavirus Police State," *VICE*, 9 April 2020, https://www.vice.com/en/article/z3bdmx/black-people-coronavirus-police-state; and Julia Sonenshein, "How Surveillance Is Changing Our Most Intimate Relationships," *New Republic*, 28 January 2024, https://newrepublic.com/article/178268/surveillance-changing-intimate-relationships.

7 Anushka Patil and Jonah Engel Bromwich, "How It Feels When Software Watches You Take Tests," *New York Times*, 29 September 2020, https://www.nytimes.com/2020/09/29/style/testing-schools-proctorio.html.

8 Ahmad and Monaghan, "World of Statues," 486.

BIBLIOGRAPHY

Ahmad, Fahad, and Jeffrey Monaghan. "Socializing the High Policing Métier: Exploring Counterterrorism Trainings for Frontline Workers." *Surveillance & Society* 20, no. 3 (2022): 248–64. https://doi.org/10.24908/ss.v20i3.14859.

Ahmad, Fahad, and Jeffrey Monaghan. "World of Statues: The 'War on Terror,' Memorialisation, and Colonial Violence." *Critical Studies on Terrorism* 14, no. 4 (2021): 484–9.

Allen, Jan. "Sorting Daemons." In *Sorting Daemons: Art, Surveillance Regimes and Social Control*, by Jan Allen, Kirsty Robertson, and Sarah EK Smith, 9–30. Kingston, ON: Agnes Etherington Art Centre, 2009.

Allen, Matthew. "See You in the City! Perth's Citiplace and the Space of Surveillance." In *Metropolis Now: Planning and the Urban in Contemporary Australia*, edited by Katherine Gibson and Sophie Watson, 137–7. Sydney: Pluto Press, 1994.

Andrejevic, Mark. "Foreword." In *Feminist Surveillance Studies*, edited by Rachel E. Dubrofsky and Shoshana A. Magnet, ix–xviii. Durham, NC: Duke University Press, 2015.

Andrews, Jolene. "Hiding or Hidden." In *Mamook Ipsoot (To Hide or Make Hidden)*. Vancouver: grunt gallery, 2013. Exhibition catalogue.

Ashcroft, Bill, Gareth Griffiths, and Helen Tiffin. *Post-Colonial Studies: The Key Concepts*. 3rd ed. London: Routledge, 2013.

Bambitchell. "Sounding the Border." Interview by Melissa Autumn White and Liz Montegary. In *Mobile Desires: The Politics and Erotics of Mobility Justice*, edited by Liz Montegary and Melissa Autumn White, 81–93. London: Palgrave Pivot, 2015.

Barnard-Wills, David, and Helen Wells. "Surveillance, Technology, and the Everyday." *Criminology and Social Justice* 12, no. 3 (2012): 227–37.

Bear Nicholas, Andrea. "The Role of Colonial Artists in the Dispossession and Displacement of the Maliseet, 1790s–1850s." *Journal of Canadian Studies* 49, no. 2 (2015): 25–86.

Benjamin, Ruha. *Race after Technology: Abolitionist Tools for the New Jim Code*. New York: Polity Press, 2019.

Bennett, Colin. "In Defense of Privacy: The Concept and the Regime." *Surveillance & Society* 8, no. 4 (2011): 485–96. https://doi.org/10.24908/ss.v8i4.4184.

Bennett, Colin J., Kevin D. Haggerty, David Lyon, and Valerie Steeves, eds. *Transparent Lives: Surveillance in Canada*. Edmonton: Athabasca University Press, 2014.

Bennett, Mia, Wilfrid Greaves, Rudolf Riedlsperger, and Alberic Botella. "Articulating the Arctic: Contrasting State and Inuit Maps of the Canadian North." *Polar Record* 52, no. 6 (2016): 630–44.

Benson, Eleanor, Morgan Brigg, Ke Hu, Sarah Maddison, Alexa Makras, Nikki Moodie, and Elizabeth Starkosch. "Mapping the Spatial Politics of Australian Settler Colonialism." *Political Geography*, no. 102 (April 2023): 1–11.

Bishop, Claire. "Participation and Spectacle: Where Are We Now?" In *Living as Form: Socially Engaged Art From 1991–2011*, edited by Nato Thompson, 34–45. New York: Creative Time Books, 2012.

Boisjoly, Raymond, and Jonah Gray. "Unsettled Sites: Haunting Canadian Colonialism." *Canadian Art*, 4 July 2016. https://canadianart.ca/reviews/unsettled-sites/.

Bouchard, David. "Observation Instruments for Imaginary Geographies." In *ISEA2014 Conference Proceedings*, edited by Thorsten Lomker, Janet Bellotto, Adina Hempel, and Woodman Taylor, 67–70. Dubai: Zayed University Books, 2015.

Bouchard, Gérard and Charles Taylor. *Building the Future: A Time for Reconciliation*. Quebec City: Government of Quebec, 2008. https://numerique.banq.qc.ca/patrimoine/details/52327/1565995.

Brighenti, Andrea Mubi. "Artveillance: At the Crossroads of Art and Surveillance." *Surveillance & Society* 7, no. 2 (2010): 137–48. https://doi.org/10.24908/ss.v7i2.4142.

Brodeur, Jean-Paul. "High Policing and Low Policing: Remarks about the Policing of Political Activities." *Social Problems* 30, no. 5 (1983): 507–20.

Browne, Simone. *Dark Matters: On the Surveillance of Blackness*. Durham, NC: Duke University Press, 2015.

Browning, Matthew, and Bruce Arrigo. "Stop and Risk: Policing, Data, and the Digital Age of Discrimination." *American Journal of Criminal Justice* 46 (2021): 298–316.

Bruno, Fernanda, Paola Barreto, and Milena Szafir. "Surveillance Aesthetics in Latin America: Work in Progress." *Surveillance & Society* 10, no. 1 (2012): 83–9.

Buenavista, Tracy Lachica, Stephanie Cariaga, Edward R. Curammeng, Elexia Reyes McGovern, Farima Pour-Khorshid, David Omotoso Stovall, and Carolina Valdez. "A Praxis of Critical Race Love: Toward the Abolition of Cisheteropatriarchy and Toxic Masculinity in Educational Justice Formations." *Education Studies* 57, no. 3 (2021): 238–49.

Buolamwini, Joy, and Timnit Gebru. "Gender Shades: Intersectional Accuracy Disparities in Commercial Gender Classification." *Proceedings of Machine Learning Research* 81 (2018): 5–19.

Burchell, Graham, Colin Gordon, and Peter Miller, eds. *The Foucault Effect: Studies in Governmentality. Two Lectures by and an Interview with Michel Foucault*. Chicago: University of Chicago Press, 1991.

Cahill, Susan. "Surveillance Frontierism." *Heliotrope*, 10 November 2021. https://www.heliotropejournal.net/helio/surveillance-frontierism.

Cahill, Susan. "Surveillance Frontierism: Art and the Colonial Project of Surveillance." *Media Practice and Education* 24, no. 2 (2023): 128–46.

Cole, Desmond. *The Skin We're In: A Year of Black Resistance and Power*. Toronto: Doubleday Canada, 2020.

Conner, Michael. "The New Normal." In *The New Normal*, edited by Independent Curators International, 12–49. New York: Artists Space, 2008. Exhibition catalogue.

Coulthard, Glen. *Red Skin, White Masks*. Minneapolis: University of Minnesota Press, 2014.

Cronon, William. "The Problem with Wilderness; Or, Getting Back to the Wrong Nature." *Environmental History* 1, no. 1 (1996): 7–28.

Crosby, Andrew, and Jeffrey Monaghan. *Policing Indigenous Movements: Dissent and the Security State*. Halifax: Fernwood Publishing, 2018.

Cyphers, Bennett, and Gennie Gebhart. *Behind the One-Way Mirror: A Deep Dive into the Technology of Corporate Surveillance*. San Francisco: Electronic Frontier Foundation, 2019.

DaViera, Andrea, Marabella Uriostegui, Aaron Gottlieb, and Ogechi (Cynthia) Onyeka. "Risk, Race, and Predictive Policing: A Critical Race Theory Analysis of the Strategic Subject List." *American Journal of Community Psychology* 73, nos. 1–2 (2023): 91–103.

Davis, Heather, and Paige Sarlin. "'On the Risk of a New Relationality': An Interview with Lauren Berlant and Michael Hardt." *Reviews in Cultural Theory* 2, no. 3 (2012). http://reviewsinculture.com/2012/10/15/on-the-risk-of-a-new-relationality-an-interview-with-lauren-berlant-and-michael-hardt/.

Deleuze, Gilles. "Postscript on the Societies of Control." *October*, no. 59 (winter 1992): 3–7.

Dewey, John. *Art as Experience*. 1934. New York: Perigee Books, 1980.

Dewey-Hagborg, Heather. "Sci-Fi Drama with a Strong, Black Lead." *New Inquiry*, 6 July 2015. https://thenewinquiry.com/sci-fi-crime-drama-with-a-strong-black-lead/.

Dickenson, Rachelle. "Processing: Ursula Johnson and the Art of Doing It Wrong." In *Mi'kwite'tmn*, 123–8. Halifax: St Mary's University Art Gallery, 2019. Exhibition catalogue. https://thereach.ca/wp-content/uploads/2021/10/PDF-Catalogue_mikwitetmn-v15.pdf.

Duarte, Daniel Edler. "The Making of Crime Predictions." *Surveillance & Society* 19, no. 2 (2021): 199–215. https://doi.org/10.24908/ss.v19i2.14261.

Edney, Matthew H. *Mapping an Empire*. Chicago: University of Chicago Press, 1997.

Egbert, Simon, and Monique Mann. "Discrimination in Predictive Policing: The (Dangerous) Myth of Impartiality and the Need for STS Analysis." In *Automating Crime Prevention, Surveillance, and Military Operations*, edited by Aleš Završnik and Vasja Badalič, 25–46. Cham, CH: Springer, 2021.

Farman, Jason. "Mapping the Digital Empire." *New Media & Society* 12, no. 6 (2010): 869–88.

Finn, Jonathan. "Seeing Surveillantly." In *Eyes Everywhere: The Global Growth of Camera Surveillance*, edited by Aaron Doyle, Randy Lippert, and David Lyon, 67–80. London: Routledge, 2011.

Finn, Jonathan. "Surveillance Studies and Visual Art: An Examination of Jill Magid's Evidence Locker." *Surveillance & Society* 10, no. 2 (2012): 134–49. https://doi.org/10.24908/ss.v10i2.4144.

Fiske, John. "Surveilling the City: Whiteness, the Black Man, and Democratic Totalitarianism." *Theory, Culture & Society* 15, no. 2 (1998): 67–88.

Fiske, John. "White Watch." In *Ethnic Minorities and the Media: Changing Cultural Boundaries*, edited by Simon Cottle, 50–66. Philadelphia: Open University Press, 2000.

Fleetwood, Nicole. *Troubling Vision: Performance, Visuality, and Blackness*. Chicago: University of Chicago Press, 2011.

Forcese, Craig, and Kent Roach. "Bill C-51: The Good, The Bad ... and the Truly Ugly." *Walrus*, 13 February 2015. http://thewalrus.ca/bill-c-51-the-good-the-bad-and-the-truly-ugly/.

Foster, Hal, ed. *Vision and Visuality*. Seattle: Bay Press, 1988.
Foucault, Michel. *Discipline and Punish: The Birth of the Prison*. New York: Pantheon, 1977.
Fung, Jessica. "Review of David Rokeby, Oakville Galleries, Oakville, Ontario, 25 June to 17 October 2004." *Ciel Variable*, no. 66 (spring 2005).
Gál, Réka Patrícia. "The Interstellar Railroad, or Speculation and Shareholder Whiteness in the Space Economy." *Heliotrope*, 14 April 2021. https://www.heliotropejournal.net/helio/the-interstellar-railroad.
Gandy, Oscar H. "Data Mining and Surveillance in the Post 9/11 Environment." In *The Surveillance Studies Reader*, edited by Sean P. Hier and Joshua Greenberg, 147–57. Berkshire, UK: McGraw-Hill Open University Press, 2007.
Gandy, Oscar. *The Panoptic Sort: A Political Economy of Personal Information*, 2nd ed. New York: Oxford University Press, 2021.
Garcia-Salazar, Paula, Nina Loshkajian, Albert Fox Cahn, and Eleni Manis. *The Spy Next Door: The Danger of Neighborhood Surveillance Apps*. New York: STOP – Surveillance Technology Oversight Project, 2021. https://www.stopspying.org/thespynextdoor#_ftn6.
Gebhard, Amanda, Sheelah McLean, and Verna St. Denis, eds. *White Benevolence: Racism and Colonial Violence in the Helping Professions*. Halifax: Fernwood Publishing, 2022.
Gibson, Ross. "Narrative Hunger: GIS Mapping, Google Street View and the Colonial Prospectus." *Cultural Studies Review* 20, no. 2 (2014): 250–65.
Gill, Rosalind. "Surveillance Is a Feminist Issue." In *The Routledge Handbook of Contemporary Feminism*, edited by Tasha Oren and Andrea Press, 148–61. London: Routledge, 2019.
Glasbeek, Amanda, Mariful Alam, and Katrin Roots. "Seeing and Not-Seeing: Race and Body-Worn Cameras in Canada." *Surveillance & Society* 18, no. 3 (2020): 328–42. https://doi.org/10.24908/ss.v18i3.13259.
Godfrey, Hannah. "Colour, Abstraction and Queerness in the Art of Derek Dunlop." *Peripheral Review*, 2 May 2020. https://www.peripheralreview.com/colour-abstraction-and-queerness-in-the-art-of-derek-dunlop/.
Goeman, Mishuana. "The Tools of a Cartographic Poet: Unmapping Settler Colonialism in Joy Harjo's Poetry." *Settler Colonial Studies* 2, no. 2 (2012): 89–112.
Goldstein, Alyosha. "On the Reproduction of Race, Capitalism, and Settler Colonialism." In *Race and Capitalism: Global Territories, Transnational Histories*, edited by Ananya Roy, 42–51. Los Angeles: UCLA Institute on Inequality and Democracy, 2018.
Gradecki, Jennifer, and Derek Curry. "Crowd-Sourced Intelligence Agency: Prototyping Counterveillance." *Big Data & Society* 4, no. 1 (2017). https://journals.sagepub.com/doi/epub/10.1177/2053951717693259.
Gregory, Derek. "From a View to a Kill: Drones and Late Modern War." *Theory, Culture & Society* 28, nos. 7–8 (2012): 188–215.
Griffiths, Michael R. "The White Gaze and Its Artifacts: Governmental Belonging and Non-Indigenous Evaluation in a (Post)-Settler Colony." *Postcolonial Studies* 15, no. 4 (2012): 415–35.
grunt gallery. *Mamook Ipsoot (To Hide or Make Hidden)*. Vancouver: grunt gallery, 2013. Exhibition catalogue.
Haggerty, Kevin D. "Foreword: Surveillance and Political Problems." In *Surveillance: Power, Problems, Politics*, edited by Sean P. Hier and Josh Greenberg, ix–xvii. Vancouver: UBC Press, 2009.
Haggerty, Kevin D., and Richard V. Ericson. "The New Politics of Surveillance and Visibility." In *The New Politics of Surveillance and Visibility*, edited by Kevin D. Haggerty and Richard V. Ericson, 3–25. Toronto: University of Toronto Press, 2005.
Haggerty, Kevin D., and Daniel Trottier. "Surveillance and/of Nature: Monitoring Beyond the Human." *Society & Animals* 23 (2015): 400–20.
Haggerty, Kevin D., Dean Wilson, and Gavin J.D. Smith. "Theorizing Surveillance in Crime Control." *Theoretical Criminology* 15, no. 3 (2011): 231–7.
Hameed, Ayesha. "Borders in the City." In *Imagining Resistance: Visual Culture and Activism in Canada*, edited by J. Keri Cronin and Kirsty Robertson, 121–39. Waterloo, ON: Wilfrid Laurier University Press, 2011.
Hier, Sean. "Probing the Surveillant Assemblage: On the Dialectics of Surveillance Practices as Processes of Social Control." *Surveillance & Society* 1, no. 3 (2003): 399–411. https://doi.org/10.24908/ss.v1i3.3347.
Hier, Sean P., and Josh Greenberg. "The Politics of Surveillance: Power, Paradigms, and the Field of Visibility." In *Surveillance: Power, Problems, Politics*, edited by Sean P. Hier and Josh Greenberg, 14–29. Vancouver: UBC Press, 2009.
hooks, bell. *Black Looks: Race and Representation*. 2nd ed. New York: Routledge, 2014.
Hopman, Roos, and Amade M'charek. "Facing the Unknown Suspect: Forensic DNA Phenotyping and the Oscillation Between the Individual and the Collective." *BioSocieties* 15 (2020): 438–62.
Ingraham, Chris, and Allison Rowland. "Performing Imperceptibility: Google Street View and the *Tableau Vivant*." *Surveillance & Society* 14, no. 2 (2016): 211–26. https://doi.org/10.24908/ss.v14i2.6013.
Janes, Robert R., and Richard Sandell, eds. *Museum Activism*. London: Routledge, 2019.

Jay, Martin, and Sumathi Ramaswamy, eds. *Empires of Vision: A Reader.* Durham, NC: Duke University Press, 2014.

Johns, Jessica. "Indigenous World-Building at the Vancouver Art Gallery." *Canadian Art*, 19 December 2019.

Johnson, Ursula. "Ursula Johnson Q&A: Of Craft and Cultural Survival." Interview by Alison Cooley. *Canadian Art*, 4 June 2014. https://canadianart.ca/interviews/ursula-johnson-qa-craft-cultural-survival/.

Kaba, Mariame. *We Do This 'Til We Free Us: Abolitionist Organizing and Transforming Justice*, edited by Tamara K. Nopper. Chicago: Haymarket Books, 2021.

Kafer, Gary. "Reimagining Resistance: Performing Transparency and Anonymity in Surveillance Art." *Surveillance & Society* 14, no. 2 (2016): 227–39.

Kaplan, Caren. *Aerial Aftermaths: Wartime from Above*. Durham, NC: Duke University Press, 2018.

Keene, Adrienne, and Gregory Hitch. "Drone Warriors: The Art of Surveillance and Resistance at Standing Rock." *Edge Effects*, 7 November 2019. https://edgeeffects.net/drone-warriors-standing-rock/.

Kim, Annabel L. "The Politics of Citation." *Diacritics* 48, no. 3 (2020): 4–9.

Kinsman, Gary, Dieter K. Buse, and Mercedes Steedman, eds. *Whose National Security? Canadian State Surveillance and the Creation of Enemies*. Toronto: Between the Lines, 2010.

Kwan, Vanessa. "About the Work." In *Antonia Hirsch: Double Blind*, edited by Public Art Vancouver. Vancouver: City of Vancouver Public Art Program, 2008. Exhibition catalogue. http://antoniahirsch.com/wp-content/uploads/2017/01/doubleblind.pdf.

Lauvin, Rémi. "Neutralizing the Surveillant Gaze: Parades of the Filmic Body." *InMedia: The French Journal of Media Studies* 8, no. 2 (2020). https://doi.org/10.4000/inmedia.2579.

Lewis, Randolph. *Under Surveillance: Being Watched in Modern America*. Austin: University of Texas Press, 2017.

Loft, Steven. "Reconstructions and Digital Landscapes." In *Constructed Land*. Toronto: InterAccess Electronic Media Arts Centre, 2012. Exhibition catalogue.

Lyon, David. "9/11, Synopticon, and Scopophilia: Watching and Being Watched." In *The New Politics of Surveillance and Visibility*, edited by Kevin D. Haggerty and Richard V. Erikson, 35–54. Toronto: University of Toronto Press, 2006.

Lyon, David. *Surveillance after September 11*. Cambridge: Polity Press, 2003.

Lyon, David. "Surveillance and the Eye of God." *Studies in Christian Ethics* 27, no. 1 (2014): 21–32.

Lyon, David. *Surveillance Society: Monitoring Everyday Life*. Buckingham, UK: Open University Press, 2001.

Lyon, David. *Surveillance Studies*. Cambridge: Polity Press, 2007.

Lyon, David. "Surveillance Technology and Surveillance Society." In *Modernity and Technology*, edited by Thomas J. Misa, Philip Brey, and Andrew Feenberg, 161–83. Cambridge, MA: MIT Press, 2003.

Lyon, David. "Under My Skin: From Identification Papers to Body Surveillance." In *Documenting Individual Identity*, edited by Jane Caplan and John Torpey, 291–310. Princeton, NJ: Princeton University Press, 2001.

Mahrouse, Gada. "'Reasonable Accommodation' in Québec: The Limits of Participation and Dialogue." *Race & Class* 52, no. 1 (2010): 85–96.

Majid, Anouar. *We Are All Moors: Ending Centuries of Crusades against Muslims and Other Minorities*. Minneapolis: University of Minnesota Press, 2009.

Matthews, Sara. "Visualizing Drone Ethnography in the Shadows of Distributive War." In *Creative Methods in Military Studies*, edited by Alice Cree, 25–40. London: Rowman and Littlefield, 2023.

Maynard, Robyn. *Policing Black Lives: State Violence in Canada from Slavery to the Present*. Halifax: Fernwood Publishing, 2017.

Maynard, Robyn, Leanne Betasamosake Simpson, Hannah Voegele, and Christopher Griffin. "Every Day We Must Get Up and Relearn the World: An Interview with Robyn Maynard and Leanne Betasamosake Simpson." *Interfere: Journal for Critical Thought and Radical Politics* 2 (November 2021): 140–65. https://interferejournal.org/wp-content/uploads/2021/11/11.-maynard-et-al.-every-day-we-must-get-up-and-relearn-the-world-2.pdf.

McDonald, Taylor Robertson. *Identity Discourses and the Canadian Foreign Policy in the War on Terror*. Cham, CH: Palgrave Macmillan, 2023.

McGrath, John Edward. *Loving Big Brother: Performance, Privacy, and Surveillance Space*. London: Routledge, 2004.

McKay, Marylin. *Picturing the Land: Narrating Territories of Canadian Landscape Art, 1500–1950*. Montreal and Kingston: McGill-Queen's University Press, 2011.

McKnight, Stéphanie. "Creative Research Methodologies for Surveillance Studies." *Surveillance & Society* 18, no. 1 (2020): 148–56. https://doi.org/10.24908/ss.v18i1.13530.

Mesquita, André. "Counter-Cartographies: The Insurrection of Maps." In *This Is Not an Atlas: A Global Collection of Counter-Cartographies*, edited by Kollektiv Orangotango+, 26–36. Bielefeld, DE: transcript Verlag, 2018.

Mirzoeff, Nicholas. "On Visuality." *Journal of Visual Culture* 5, no. 1 (2006): 53–79.

Mirzoeff, Nicholas. “The Right to Look.” *Critical Inquiry* 37, no. 3 (2011): 473–96.

Mirzoeff, Nicholas. *The Right to Look: A Counterhistory of Visuality.* Durham, NC: Duke University Press, 2011.

Mitchell, W.J.T. “Showing Seeing.” *Journal of Visual Culture* 1, no. 2 (2002): 165–81.

Mitra, Srimoyee, ed. *Border Cultures*. London: Black Dog Publishing, 2015. Exhibition catalogue.

Monaghan, Jeffrey. “Security Traps and Discourses of Radicalization: Examining Security Practices Targeting Muslims in Canada.” *Surveillance & Society* 12, no. 4 (2014): 485–501. https://doi.org/10.24908/ss.v12i4.4557.

Monaghan, Jeffrey. “Settler Governmentality and Racializing Surveillance in Canada’s North-West.” *Canadian Journal of Sociology*, 38, no. 4 (2013): 487–508.

Monaghan, Jeffrey, and Andrew Crosby. *Policing Indigenous Movements: Dissent and the Security State*. Halifax: Fernwood Publishing, 2018.

Monahan, Torin. *Crisis Vision: Race and Cultural Production of Surveillance*. Durham, NC: Duke University Press, 2022.

Monahan, Torin. “The Right to Hide? Anti-Surveillance Camouflage and the Aesthetics of Resistance.” *Communication and Critical/Cultural Studies* 12, no. 2 (2015): 159–78.

Monahan, Torin. “Ways of Being Seen: Surveillance Art and the Interpellation of Viewing Subjects.” *Cultural Studies* 32, no. 4 (2018): 560–81.

Moreton-Robinson, Aileen. *White Possessive: Property, Power, and Indigenous Sovereignty*. Minneapolis: University of Minnesota Press, 2015.

Morrison, Elise. *Discipline and Desire: Surveillance Technologies in Performance*. Ann Arbor: University of Michigan Press, 2016.

Morton, Erin. “Introduction: Unsettling Canadian Art History.” In *Unsetting Canadian Art History*, edited by Erin Morton, 1–42. Montreal and Kingston: McGill-Queen’s University Press, 2022.

Moser, Gabrielle. “Jennifer Marman and David Borins.” *Canadian Art*, 15 March 2013. https://canadianart.ca/reviews/jennifer-marmans-and-daniel-borins/.

Murray-Lichtman, Andrea, and Siham Elkassem. “Academic Voyeurism: The White Gaze in Social Work.” *Canadian Social Work Review* 38, no. 2 (2021): 179–205.

Nakamura, Lisa. “Afterword: Blaming, Shaming and the Feminization of Social Media.” In *Feminist Surveillance Studies*, edited by Rachel E. Dubrofsky and Shoshana Amielle Magnet, 221–28. Durham, NC: Duke University Press, 2015.

Newell, Bryce Clayton. “Context, Visibility, and Control: Police Work and the Contested Objectivity of Bystander Video,” *New Media & Society* 21, no. 1 (2019): 60–76.

Noble, Safiya Umoja. *Algorithms of Oppression: How Search Engines Reinforce Racism*. New York: New York University Press, 2018.

Nunn, Neil. “Toxic Encounters, Settler Logics of Elimination, and the Future of a Continent.” *Antipode* 50, no. 5 (2018): 1330–48. https://doi.org/10.1111/anti.12403.

Ogasawara, Midori. “Mainstreaming Colonial Experiences in Surveillance Studies.” *Surveillance & Society* 17, no. 5 (2019): 726–9. https://doi.org/10.24908/ss.v17i5.13521.

O’Neill, Paul. *The Culture of Curating and the Curating of Culture(s).* Cambridge, MA: MIT Press, 2016.

Ontario Human Rights Commission. “Stop and Search: Gaps in Regulations, Policies, and Procedures.” Chapter 5 in *From Impact to Action: Final Report into Anti-Black Racism by the Toronto Police Service*. Ontario Human Rights Commission, 2023. https://www.ohrc.on.ca/en/impact-action-final-report-anti-black-racism-toronto-police-service/chapter-5-stop-and-search-gaps-regulations-policies-and-procedures.

Ożóg, Maciej. “Surveilling the Surveillance Society: The Case of Rafael Lozano-Hemmer’s Installations.” In *Conspiracy Dwellings: Surveillance in Contemporary Art*, edited by Outi Remes and Pam Skelton, 95–112. Newcastle upon Tyne: Cambridge Scholars Publishing, 2010.

Page, Allison, and Carmen Jones. “Weaponizing Neutrality: The Entanglement of Police, Affect, and Surveillance Technologies.” *Feminist Media Studies* 23, no. 1 (2023): 170–84.

Patel, Shaista. “The Anti-Terrorism Act and National Security: Safeguarding the Nation against Uncivilized Muslims.” In *Islam in the Hinterlands: Muslim Cultural Politics in Canada*, edited by Jasmine Zine, 272–98. Vancouver: UBC Press, 2012.

Patel, Shaista. “Talking Complicity, Breathing Coloniality: Interrogating Settler-centric Pedagogy of Teaching about White Settler Colonialism.” *Journal of Curriculum and Pedagogy* 19, no. 3 (2022): 211–30.

Patel, Shaista, Ghaida Moussa, and Nishant Upadhyay. “Complicities, Connections and Struggles: Critical Transnational Feminist Analysis of Settler Colonialism.” *Feral Feminisms* 4 (summer 2015): 5–19.

Proulx, Craig. “Colonizing Surveillance: Canada Constructs an Indigenous Terror Threat.” *Anthropologica* 56, no. 1 (2014): 83–100.

Puar, Jaspir. *Terrorist Assemblages: Homonationalism in Queer Times*. Durham, NC: Duke University Press, 2007.

Queirós, Filipa. “The Visibilities and Invisibilities of Race Entangled with Forensic DNA Phenotyping Technology.” *Journal of Forensic and*

Legal Medicine 68 (November 2019): 1–7. https://www.sciencedirect.com/science/article/pii/S1752928X19300873.

Raji, Inioluwa Deborah, Timnit Gebru, Margaret Mitchell, Joy Buolamwini, Joonseok Lee, and

Emily Denton. "Saving Face: Investigating the Ethical Concerns of Facial Recognition Auditing." In *Proceedings of the 2020 AAAI/ACM Conference on AI, Ethics, and Society (AIES '20)*, 145–51. New York: AAAI/ACM Conference, 2020. https://doi.org/10.1145/3375627.3375820.

Rice, Carla, Susan D. Dion, Hannah Fowlie, and Andrea Breen. "Re/turning the Gaze: Unsettling Settler Logics through Multimedia Storytelling." *Feminist Media Studies* 22, no. 3 (2022): 638–56.

Robé, Chris. *Abolishing Surveillance: Digital Media Activism and State Repression*. New York: PM Press, 2023.

Robertson, Carmen. "Notes to a Nation: Teachings on Land through the Art of Norval Morrisseau." In *Unsettling Canadian Art History*, edited by Erin Morton, 65–86. Montreal and Kingston: McGill-Queen's University Press, 2022.

Robertson, Kirsty. *Tear Gas Epiphanies: Protest, Culture, Museums*. Montreal and Kingston: McGill-Queen's University Press, 2019.

Robertson, Kirsty. "'Try to Walk with the Sound of My Footsteps': The Surveillant Body in Contemporary Art." *Communication Review* 11 (January–March 2008): 24–41.

Rogoff, Irit. "Smuggling: An Embodied Criticality." *EIPCP, European Institute for Progressive Politics* (2006). https://eipcp.net/dlfiles/rogoff-smuggling/attachment download/rogoff-smuggling.pdf.

Rudd, Judith, and Michèle Sedgwick, eds. *Issues in Curating Contemporary Art and Performance*. Chicago: Chicago University Press, 2007.

Sa'di, Ahmad H. "Colonialism and Surveillance." In *Routledge Handbook of Surveillance Studies*, edited by Kirstie Ball, Kevin D. Haggerty, and David Lyon, 151–8. London: Routledge, 2012.

Shapiro, Aaron. "Predictive Policing for Reform? Indeterminacy and Intervention in Big Brother Data." *Surveillance & Society* 17, nos. 3–4 (2019): 456–72. https://doi.org/10.24908/ss.v17i3/4.10410.

Sharma, Nandita. "Racism." In *Citizenship and Its Others*, edited by Bridget Anderson and Vanessa Hughes, 98–118. London: Palgrave Macmillan, 2015.

Sharpe, Christina. "Variations in Black Queer Worlds." *Canadian Art*, 20 August 2019. https://canadianart.ca/essays/variations-in-black-queer-worlds/.

Sherman, Stephanie. "The Polyopticon: A Diagram for Urban Intelligence." *AI & Society* 38 (2023): 1209–22.

Sholette, Gregory. *The Art of Activism and the Activism of Art*. Chicago: Lund Humphries, 2021.

Simpson, Audra. "The State Is a Man." *Theory & Event* 19, no. 4 (2016). https://muse.jhu.edu/article/633280.

Simpson, Audra. "Whither Settler Colonialism?" *Settler Colonial Studies* 6, no. 4 (2016): 438–45.

Smith, David R. "Sociable Laughter, Deep Laughter." In *Parody and Festivity in Early Modern Art: Essays on Comedy as Social Vision*, edited by David R. Smith, 1–10. Burlington, VT: Ashgate Publishing, 2012.

Smith, Gavin J.D. "Surveillance, Data, and Embodiment: On the Work of Being Watched." *Body & Society* 22, no. 2 (2016): 108–39.

Smith, Keith D. *Liberalism, Surveillance, and Resistance: Indigenous Communities in Western Canada, 1877–1927*. Athabasca, AB: Athabasca University Press, 2009.

Sorkin, Michael, ed. *Indefensible Space: The Architecture of the National Insecurity State*. New York: Routledge, 2008.

Stahl, Roger, and Sebastian Kaempf. "Sousveilling the 'Global War on Terror.'" *Australian Journal of International Affairs* 73, no. 4 (2019): 337–56.

Stevens, Quentin, and Julia Lossau. "Framing Art and Its Uses in Public Spaces." In *The Uses of Art in Public Spaces*, edited by Quentin Stevens and Julia Lossau, 1–16. New York: Routledge, 2015.

Stroud, Matthew. "The Most Advanced Police Sketch Ever Might Solve Cold Cases." *The Verge*, 20 July 2014. https://www.theverge.com/2014/7/20/5916661/the-most-advanced-police-sketch-ever-might-solve-cold-cases.

Sturken, Marita, and Lisa Cartwright. *Practices of Looking: An Introduction to Visual Culture*. New York: Oxford University Press, 2009.

Tremblay, Élène. "*Stranger Visions* by Heather Dewey-Hagborg: Reinterpreting Portraiture Through New Forensic and 3D Printing Techniques." *ETC Media* no. 103 (2014–15): 44–8.

Tuck, Eve, and Wayne Yang. "Decolonization Is Not a Metaphor." *Decolonization: Indigeneity, Education & Society* 1, no. 1 (2012): 1–40.

Valadares, Desiree. "Uneven Mobilities: Infrastructural Imaginaries on the Hope-Princeton Highway." *Radical History Review* 147 (October 2023): 158–85.

van der Meulen, Emily, and Robert Heynan, eds. *Expanding the Gaze: Gender and the Politics of Surveillance*. Toronto: University of Toronto Press, 2016.

van der Ploeg, Irma. "The Body as Data in the Age of Information." In *Routledge Handbook of Surveillance Studies*, edited by Kirstie Ball, Kevin Haggerty, and David Lyon, 176–83. London: Routledge, 2012.

Vukov, Tamara, and Mimi Sheller. "Border Work: Surveillant Assemblages, Virtual Fences, And Tactical Counter-Media." *Social Semiotics* 23, no. 2 (2013): 225–41.

Wakefield, Allison. “The Public Surveillance Functions of Private Spaces.” *Surveillance & Society* 2, no. 4 (2004): 529–45. https://doi.org/10.24908/ss.v2i4.3362.

Waldron, Ingrid. “Mapping Racial Geographies of Violence on the Colonial Landscape.” *Windsor Yearbook of Access to Justice* 38 (2022): 1–9.

Walker, Ellyn. “Variations in Black, Queer, and Otherwise: Works by Adbi Osman.” Toronto: Art Museum at the University of Toronto, 2019. Exhibition catalogue. https://artmuseum.utoronto.ca/wp-content/uploads/2019/04/2019_Osman-Brochure-r2.pdf.

Wallace, Belinda Doneen. “Our Lives: Scribal Activism, Intimacy, and Black Lesbian Visibility in 1980s Canada.” *Journal of Canadian Studies* 54, nos. 2–3 (2020): 334–59.

Weinberg, Lindsay. “Surveillance as Paradox: A Review of *Feminist Surveillance Studies*.” *Studies in Control Societies* 2, no. 1 (2017). https://studiesincontrolsocieties.org/surveillance-as-paradox/.

Willett, Graham, and Yorick Smaal, eds. *Intimacy, Violence and Activism: Gay and Lesbian Perspectives on Australasian History and Society*. Clayton, AU: Monash University Publishing, 2011.

Wolfe, Patrick. “Settler Colonialism and the Elimination of the Native.” *Journal of Genocide Research* 8, no. 4 (2006): 387–409.

Wood, David. “Foucault and Panopticonism Revisited.” *Surveillance & Society* 1, no. 3 (2003): 234–9. https://doi.org/10.24908/ss.v1i3.3338.

Yar, Majid. “Panoptic Power and the Pathologisation of Vision: Critical Reflections on the Foucauldian Thesis.” *Surveillance & Society* 1, no. 3 (2003): 254–71. https://doi.org/10.24908/ss.v1i3.3340.

Zine, Jasmin. *Under Siege: Islamophobia and the 9/11 Generation*. Montreal and Kingston: McGill-Queen’s University Press, 2022.

Zureik, Elia, L. Lynda Harling Stalker, Emily Smith, David Lyon, and Yolande E. Chan, eds. *Surveillance, Privacy, and the Globalization of Personal Information: International Comparisons*. Montreal and Kingston: McGill-Queen’s University Press, 2010.

INDEX

McGill-Queen's/Beaverbrook Canadian Foundation Studies in Art History

Martha Langford and Erin Silver, series editors

The Practice of Her Profession
Florence Carlyle, Canadian Painter in the Age of Impressionism
Susan Butlin

Bringing Art to Life
A Biography of Alan Jarvis
Andrew Horrall

Picturing the Land
Narrating Territories in Canadian Landscape Art, 1500 to 1950
Marylin J. McKay

The Cultural Work of Photography in Canada
Edited by Carol Payne and Andrea Kunard

Newfoundland Modern
Architecture in the Smallwood Years, 1949–1972
Robert Mellin

The Codex Canadensis *and the Writings of Louis Nicolas*
The Natural History of the New World, Histoire Naturelle des Indes Occidentales
Edited and with an Introduction by François-Marc Gagnon, Translation by Nancy Senior, Modernization by Réal Ouellet

Museum Pieces
Toward the Indigenization of Canadian Museums
Ruth B. Phillips

The Allied Arts
Architecture and Craft in Postwar Canada
Sandra Alfoldy

Rethinking Professionalism
Essays on Women and Art in Canada, 1850–1970
Edited by Kristina Huneault and Janice Anderson

The Official Picture
The National Film Board of Canada's Still Photography Division and the Image of Canada, 1941–1971
Carol Payne

Paul-Émile Borduas
A Critical Biography
François-Marc Gagnon
Translated by Peter Feldstein

On Architecture
Melvin Charney: A Critical Anthology
Edited by Louis Martin

Making Toronto Modern
Architecture and Design, 1895–1975
Christopher Armstrong

Negotiations in a Vacant Lot
Studying the Visual in Canada
Edited by Lynda Jessup, Erin Morton, and Kirsty Robertson

Visibly Canadian
Imaging Collective Identities in the Canadas, 1820–1910
Karen Stanworth

Breaking and Entering
The Contemporary House Cut, Spliced, and Haunted
Edited by Bridget Elliott

Family Ties
Living History in Canadian House Museums
Andrea Terry

Picturing Toronto
Photography and the Making of a Modern City
Sarah Bassnett

Architecture on Ice
A History of the Hockey Arena
Howard Shubert

For Folk's Sake
Art and Economy in Twentieth-Century Nova Scotia
Erin Morton

Spaces and Places for Art
Making Art Institutions in Western Canada, 1912–1990
Anne Whitelaw

Narratives Unfolding
National Art Histories in an Unfinished World
Edited by Martha Langford

Canadian Painters in a Modern World, 1925–1955
Writings and Reconsiderations
Lora Senechal Carney

Sketches from an Unquiet Country
Canadian Graphic Satire, 1840–1940
Edited by Dominic Hardy, Annie Gérin, and Lora Senechal Carney

I'm Not Myself at All
Women, Art, and Subjectivity in Canada
Kristina Huneault

The Global Flows of Early Scottish Photography
Encounters in Scotland, Canada, and China
Anthony W. Lee

Tear Gas Epiphanies
Protest, Culture, Museums
Kirsty Robertson

What Was History Painting and What Is It Now?
Edited by Mark Salber Phillips and Jordan Bear

Through Post-Atomic Eyes
Edited by Claudette Lauzon and John O'Brian

Jean Paul Riopelle et le mouvement automatiste
François-Marc Gagnon

Jean Paul Riopelle and the Automatiste Movement
François-Marc Gagnon
Translated by Donald Winkler

I Can Only Paint
The Story of Battlefield Artist Mary Riter Hamilton
Irene Gammel

Object Lives and Global Histories in Northern North America
Material Culture in Motion, c. 1780–1980
Edited by Beverly Lemire, Laura Peers, and Anne Whitelaw

For the Temporary Accommodation of Settlers
Architecture and Immigrant Reception in Canada, 1870–1930
David Monteyne

Women at the Helm
How Jean Sutherland Boggs, Hsio-yen Shih, and Shirley L. Thomson Changed the National Gallery of Canada
Diana Nemiroff

Voluntary Detours
Small-Town and Rural Museums in Alberta
Lianne McTavish

Photogenic Montreal
Activisms and Archives in a Post-industrial City
Edited by Martha Langford and Johanne Sloan

Wendat Women's Arts
Annette W. de Stecher

Unsettling Canadian Art History
Edited by Erin Morton

Out of School
Information Art and the Toronto School of Communication
Adam Lauder

Jackson's Wars
A.Y. Jackson, the Birth of the Group of Seven, and the Great War
Douglas Hunter

Out of the Studio
The Photographic Innovations of Charles and John Smeaton at Home and Abroad
John Osborne and Peter Smeaton

Casa Loma
Millionaires, Medievalism, and Modernity in Toronto's Gilded Age
Edited by Matthew M. Reeve and Michael Windover

Variable Conditions
Para-computational Arts in Canada, 1965–1995
Edited by Adam Lauder

Needle Work
A History of Commercial Tattooing in Canada
Jamie Jelinski

A History of Photography in Canada
Volume 1: Anticipation to Participation, 1839–1918
Martha Langford

Toronto Edwardian
Frank Darling, Architect of Canada's Imperial Age
David E. Winterton

Classicism in Canada
Ambition, Utopia, Hubris
Edited by Joan Coutu with David A. Galbraith

The Hunter, the Crown, and the Cameras
Joseph Idlout and the Image of Sangussaqtauliqtilluta
Carol Payne
With texts and images by Joseph Idlout and Leah Idlout d'Argencourt Paulson
With contributions by Madeleine d'Argencourt and Augatnaaq Eccles
And with the words of Paul Ullatitaq Idlout, Susan Salluviniq, Moses Idlout, Joshua Idlout, Ludy Pudluk, Jacob Anaviapik, Elisapee Ootoova, Ruth Sangoya, and Solomon Awa

States of Observance
Contemporary Art and Surveillance Logic in Canada
Susan Cahill